D1346026

Jeffrey T. Huber, PhD
Editor

HIV/AIDS
Internet Information
Sources and Resources

HIV/AIDS Internet Information Sources and Resources has been co-published simultaneously as *Health Care on the Internet*, Volume 2, Numbers 2/3 1998.

Pre-publication
REVIEWS,
COMMENTARIES,
EVALUATIONS . . .

" **T**his guidebook will demystify and detangle the World Wide Web for the AIDS researcher, the primary care provider, and the patient anxious for first-hand information in various levels of detail . . . right from the source.

Confronted by information overload the seeker of AIDS information will celebrate this handbook that covers the width and breadth of the field."

Kiyoshi Kuromiya
Director, Critical Path AIDS Project

"**W**ith the burgeoning sea of information available on the Internet, this book serves as a lighthouse for those seeking AIDS/HIV-related resources. Although there is a proliferation of information now available on the Internet, finding exactly what you need is often an exasperating process. This resource not only guides the reader to resources related to a particular need, but also perseveres to explain the purpose and scope of the information or organization. The authors impart the historical imperative of information in the continuing endeavor to treat, to find a cure, and to prevent HIV infection. As one integrated resource, this book offers an open door to authoritative Internet sites relating to treatment, research, support, service organizations, and population-specific areas of interest. Information professionals and health care providers will find this book a valuable tool for not only discovering new and important resources, but in building, evaluating, and maintaining local Internet resources on HIV and AIDS. At last! One source that directs readers to the broad spectrum of AIDS/HIV-related information resources."

Jana C. Allcock, MLS
*Coordinator for Access to Electronic Information Services
University of Arkansas
for Medical Sciences Library*

"**H**IV/AIDS Internet Information Sources and Resources* is the most current and comprehensive reference material available to assist users (both the health and allied health professionals as well as the general public) in manipulating the wealth of information on the Internet. This volume serves as a direct pointer to various Internet locations on HIV/AIDS-related information. The editor and the contributors provide a well-balanced, perceptive, and non-discriminating reference for people who need to access information concerning HIV/AIDS on the Internet.

HIV/AIDS Internet Information Sources and Resources will be a primary resource and a useful tool in identifying and facilitating access to HIV/AIDS-related Internet information for all people with Internet access world-wide. In addition, resources covered in this volume provide social support and information on quality of life for those individuals, such as HIV positive people and professional care-providers who desperately seek compassion and the knowledge for achieving miracles."

Dr. Feili B. Tu
*Visiting Assistant Professor
School of Library and Information Science
The University of Iowa*

HIV/AIDS Internet Information Sources and Resources

HIV/AIDS Internet Information Sources and Resources has been co-published simultaneously as *Health Care on the Internet*, Volume 2, Numbers 2/3 1998.

The *Health Care on the Internet* Monographs/"Separates"

Cancer Resources on the Internet, edited by M. Sandra Wood and Eric P. Delozier

HIV/AIDS Internet Information Sources and Resources, edited by Jeffrey T. Huber

These books were published simultaneously as special thematic issues of *Health Care on the Internet* and are available bound separately. Visit Haworth's website at http://www.haworthpressinc.com to search our online catalog for complete tables of contents and ordering information for these and other publications. Or call 1-800-HAWORTH (outside US/Canada: 607-722-5857), Fax: 1-800-895-0582 (outside US/Canada: 607-771-0012), or e-mail: getinfo@haworthpressinc.com

HIV/AIDS Internet Information Sources and Resources

Jeffrey T. Huber, PhD
Editor

HIV/AIDS Internet Information Sources and Resources has been co-published simultaneously as *Health Care on the Internet*, Volume 2, Numbers 2/3 1998.

The Haworth Press, Inc.
New York • London

HIV/AIDS Internet Information Sources and Resources has been co-published simultaneously as *Health Care on the Internet*™, Volume 2, Numbers 2/3 1998.

The Haworth Press, Inc., 10 Alice Street, Binghamton, NY 13904-1580 USA

Cover design by Jennifer M. Gaska

Library of Congress Cataloging-in-Publication Data

HIV/AIDS Internet information sources and resources / Jeffrey T. Huber, editor.
    p. cm.
    "Has also been co-published simultaneously as Health care on the Internet, Volume 2, Numbers 2/3, 1998."
    Includes bibliographical references and index.
    ISBN 0-7890-0544-1 (alk. paper).–ISBN 1-56023-117-3 (alk. paper)
    1. AIDS (Disease)–Computer network resources. I. Huber, Jeffrey T. II. Health care on the Internet.
RA644.A25H575526 1998
025.06′3621′969792–dc21
                                            98-8526
                                              CIP

INDEXING & ABSTRACTING

Contributions to this publication are selectively indexed or abstracted in print, electronic, online, or CD-ROM version(s) of the reference tools and information services listed below. This list is current as of the copyright date of this publication. See the end of this section for additional notes.

- *Adis International Ltd.*, 41 Centurion Drive, Mairangi Bay, North Shore City, New Zealand

- *AgeLine Database,* American Association of Retired Persons, 601 E Street NW, Washington, DC 20049

- *Applied Social Sciences Index & Abstracts (ASSIA) (Online: ASSI via Data-Star) (CDRom: ASSIA Plus),* Bowker-Saur Limited, Maypole House, Maypole Road, East Grinstead, West Sussex RH19 1HH, England

- *Brown University Digest of Addiction Theory and Application, The (DATA Newsletter),* Project Cork Institute, Dartmouth Medical School, 14 South Main Street, Suite 2F, Hanover, NH 03755-2015

- *Cambridge Scientific Abstracts, Health & Safety Science Abstracts,* 7200 Wisconsin Avenue #601, Bethesda, MD 20814

- *CINAHL (Cumulative Index to Nursing & Allied Health Literature, in print, also on CD-ROM from CD Plus, EBSCO, and SilverPlatter, and online from CDP Online (formerly BRS), Data-Star, and PaperChase. (Support materials include Subject Heading List, Database Search Guide, and instructional video.)* CINAHL Information Systems, P.O. Box 871/1509 Wilson Terrace, Glendale, CA 91209-0871

- *CNPIEC Reference Guide: Chinese National Directory of Foreign Periodicals*, P.O. Box 88, Beijing, People's Republic of China

- *Combined Health Information Database (CHID)*, National Institutes of Health, 3 Information Way, Bethesda, MD 20892-3580

(continued)

- *Computing Reviews,* Association for Computing Machinery, 1515 Broadway, 17th Floor, New York, NY 10036

- *Current Awareness Abstracts,* Association for Information Management, Information House, 20-24 Old Street, London EC1V 9AP England

- *European Association for Health Information & Libraries: selected abstracts in newsletter "Publications" section,* EAHIL Newsletter, Via Piranesi 38, I-20137 Milano, Italy

- *Health Care Literature Information Network-HECLINET,* Technische Universitat Berlin-Dokumentation Krankenhauswesen, Sekr. A42, Strasse des 17. Juni 135, D 10623 Berlin, Germany

- *Health Service Abstracts (HSA),* Department of Health, Quarry House, Room 5C07, Quarry Hill, Leeds, LS2 7UE United Kingdom

- *Healthcare Leadership Review,* COR Healthcare Resources, P.O. Box 40959, Santa Barbara, CA 93140

- *Index to Periodical Articles Related to Law,* University of Texas, 727 East 26th Street, Austin, TX 78705

- *Information Science Abstracts,* Information Today, Inc., Department ISA, 143 Old Marlton Pike, Medford, NJ 08055-8750

- *INSPEC Information Services,* Institution of Electrical Engineers, Michael Faraday House, Six Hills Way, Stevenage Herts SG1 2AY, England

- *INTERNET ACCESS (& additional networks) Bulletin Board for Libraries ("BUBL") coverage of information resources on INTERNET, JANET, and other networks.*
  - <URL:http://bubl.ac.uk/>
  - The new locations will be found under <URL:http://bubl.ac.uk/link/>.
  - Any existing BUBL users who have problems finding information on the new service should contact the BUBL help line by sending e-mail to <bubl@bubl.ac.uk>.
  The Andersonian Library, Curran Building, 101 St. James Road, Glasgow G4 0NS, Scotland

(continued)

- *Journal of the American Dietetic Association (Abstract Section),* The American Dietetic Association, 216 West Jackson Boulevard, Suite 800, Chicago, IL 60606-6995

- *Leeds Medical Information,* University of Leeds, Leeds LS2 9JT United Kingdom

- *Library & Information Science Abstracts (LISA),* Bowker-Saur Limited, Maypole House, Maypole Road, East Grinstead, West Sussex, RH19 1HH, England

- *Medicinal & Aromatic Plants Abstracts (MAPA),* Publications & Information Directorate, Hillside Road, New Delhi-110 012, India

- *OT Bibsys,* American Occupational Therapy Foundation, P.O. Box 31220, Rockville, MD 20824-1220

- *PASCAL, c/o Insititute de L'Information Scientifique et Technique,* Cross-disciplinary electronic database covering the fields of science, technology & medicine. Also available on CD-ROM, and can generate customized retrospective searches. For more information: INIST, Customer Desk, 2, allee du Parc de Brabois, F-54514 Vandoeuvre Cedex, France; http//www.inist.fr

- *Patient Care Management Abstracts,* COR Healthcare Resources, P.O. Box 40959, Santa Barbara, CA 93140-0959

- *Pharmacy Business,* Southeastern University, 3200 South University Drive, Ft. Lauderdale, FL 33328-2018

- *Referativnyi Zhurnal (Abstracts Journal of the All-Russian Institute of Scientific and Technical Information),* 20 Usievich Street, Moscow 125219, Russia

- *Social Work Abstracts,* National Association of Social Workers, 750 First Street NW, 8th Floor, Washington, DC 20002

(continued)

SPECIAL BIBLIOGRAPHIC NOTES

related to special journal issues (separates)
and indexing/abstracting

☐ indexing/abstracting services in this list will also cover material in any "separate" that is co-published simultaneously with Haworth's special thematic journal issue or DocuSerial. Indexing/abstracting usually covers material at the article/chapter level.

☐ monographic co-editions are intended for either non-subscribers or libraries which intend to purchase a second copy for their circulating collections.

☐ monographic co-editions are reported to all jobbers/wholesalers/approval plans. The source journal is listed as the "series" to assist the prevention of duplicate purchasing in the same manner utilized for books-in-series.

☐ to facilitate user/access services all indexing/abstracting services are encouraged to utilize the co-indexing entry note indicated at the bottom of the first page of each article/chapter/contribution.

☐ this is intended to assist a library user of any reference tool (whether print, electronic, online, or CD-ROM) to locate the monographic version if the library has purchased this version but not a subscription to the source journal.

☐ individual articles/chapters in any Haworth publication are also available through the Haworth Document Delivery Service (HDDS).

HIV/AIDS Internet Information Sources and Resources

CONTENTS

ABOUT THE EDITOR

Jeffrey T. Huber, PhD, is Research Information Scientist in the Active Digital Library of the Eskind Biomedical Library at Vanderbilt University in Nashville, Tennessee. He is also Research Assistant Professor in the Division of Biomedical Informatics at Vanderbilt. For two years, Dr. Huber was Assistant Professor at the School of Library and Information Studies at Texas Woman's University in Denton, Texas, where he was also Co-Chair of the University's Task Force on HIV/AIDS. He is the author of *HIV/AIDS Community Information Services: Experiences in Serving Both At-Risk and HIV-Infected Populations* (The Haworth Press, Inc.) and the co-author of *HIV/AIDS and HIV/AIDS-Related Terminology: A Means of Organizing the Body of Knowledge* (The Haworth Press, Inc.). The editor of *How to Find Information About AIDS, Second Edition* and of *Dictionary of AIDS-Related Terminology,* he has written a number of articles concerning AIDS information and has delivered many presentations at national conferences and meetings.

Introduction

Jeffrey T. Huber

Infection with the human immunodeficiency virus (HIV) results in a complex, chronic disease process, complicated by myriad economic, legal, religious, psychological, social, and spiritual factors. This chronic disease is characterized by a varied cluster of signs and symptoms that typically progress to a diagnosis of acquired immune deficiency syndrome (AIDS). HIV is differentiated from other chronic disease processes by the variety of cancers and opportunistic infections commonly associated with AIDS, as well as HIV-related dementia and wasting, and the wide variation in disease course progression and dying trajectory.

Information has been, and continues to be, viewed as a key resource in: preventing infection with the human immunodeficiency virus, managing various medical complications associated with the disease, assisting with non-biomedical complexities, and ultimately extending life expectancy. HIV/AIDS-related information forms the building blocks for education and prevention programs, treatment advances, social support, and coping mechanisms. In fact, information is the initial step in the HIV/AIDS communication continuum. This communication model, though, is marred by complexities similar to those of the disease itself. Locating and accessing timely, relevant information often proves to be a challenge. HIV/AIDS-related information is produced by many individuals, communities, organizations, and institutions in most every format available. Further complicating the situation, the AIDS epidemic has yielded its own vocabulary, one that is representative of the diverse groups of individuals infected with the virus and those working to stop the spread of the disease and ultimately find a cure.

[Haworth co-indexing entry note]: "Introduction." Huber, Jeffrey T. Co-published simultaneously in *Health Care on the Internet* (The Haworth Press, Inc.) Vol. 2, No. 2/3, 1998, pp. 1-2; and: *HIV/AIDS Internet Information Sources and Resources* (ed: Jeffrey T. Huber) The Haworth Press, Inc., 1998, pp. 1-2; and: *HIV/AIDS Internet Information Sources and Resources* (ed: Jeffrey T. Huber) Harrington Park Press, an imprint of The Haworth Press, Inc., 1998, pp. 1-2. Single or multiple copies of this article are available for a fee from The Haworth Document Delivery Service [1-800-342-9678, 9:00 a.m. - 5:00 p.m. (EST). E-mail address: getinfo@haworthpressinc.com].

1

Currently, there is a tremendous amount of HIV/AIDS-related information available, with much of that information being accessible through the Internet and World-Wide Web. Following is a selection of articles designed to highlight HIV/AIDS-specific Internet information sources and resources. The selection reflects the breadth and depth of information available, as well as issues surrounding developing and maintaining a Web presence, evaluating Internet sites, and locating relevant, reliable HIV/AIDS information.

The AIDS Community-Based Service Organization and the World-Wide Web: Decisions and Experiences in Creating a Web Site

Sheree Huber Williams

SUMMARY. Education and information dissemination are integral to the mission of AIDS community-based service organizations across the country. The World-Wide Web provides a different vehicle for fulfilling this mission. For many of these organizations, creating a Web site will be a new and challenging endeavor. This article raises some of the issues that should be addressed at the planning stages and discusses the experience of Louisville's Community Health Trust in beginning the Web site creation process. *[Article copies available for a fee from The Haworth Document Delivery Service: 1-800-342-9678. E-mail address: getinfo@haworthpressinc.com]*

AIDS community-based service organizations across the country have incorporated education and information dissemination into their mission statements. With the general public's growing access to the World-Wide

Sheree Huber Williams (swilliam@pop.jcc.uky.edu) is Director of Library Services at Jefferson Community College, Downtown Campus, in Louisville, Kentucky. She is a volunteer at the Community Health Trust, the AIDS community-based service organization in Louisville. The author extends special thanks to Dr. Emery Lane, President of the Community Health Trust Board of Directors.

[Haworth co-indexing entry note]: "The AIDS Community-Based Service Organization and the World-Wide Web: Decisions and Experiences in Creating a Web Site." Williams, Sheree Huber. Co-published simultaneously in *Health Care on the Internet* (The Haworth Press, Inc.) Vol. 2, No. 2/3, 1998, pp. 3-9; and: *HIV/AIDS Internet Information Sources and Resources* (ed: Jeffrey T. Huber) The Haworth Press, Inc., 1998, pp. 3-9; and: *HIV/AIDS Internet Information Sources and Resources* (ed: Jeffrey T. Huber) Harrington Park Press, an imprint of The Haworth Press, Inc., 1998, pp. 3-9. Single or multiple copies of this article are available for a fee from The Haworth Document Delivery Service [1-800-342-9678, 9:00 a.m. - 5:00 p.m. (EST). E-mail address: getinfo@haworthpressinc.com].

Web, these organizations are adding a presence on the Web to the list of venues used to educate and inform their communities. The ability to access information from the privacy of one's home makes this source particularly appealing for some. Because of the nature of community-based service organizations, the creation of that Web presence can be daunting. Funding, staffing, and technical issues must all be considered in the process. The Community Health Trust in Louisville, Kentucky, is in the early stages of Web site development and has encountered some hurdles along the way. Approaching this process logically and planning carefully should ease some frustration and result in a Web site that works for your community.

One of the first and most important questions to ask once your organization has decided to create a Web site is who will be responsible for it. Limited staffing is common in community-based service organizations; volunteers, therefore, may become a critical element in the "who" question. In fortunate circumstances, a staff member can absorb this task. More likely, a staff member, perhaps the education coordinator, since most AIDS service organizations do not employ information professionals,[1] will coordinate the effort with a volunteer or group of volunteers. Often involvement will be determined by a staff member's experience or interest in the Web rather than by where the project may fall logically within the organizational structure. In some instances, no staff time will be available, and the project will be dependent on a volunteer effort.

Responsibility must be determined for decision-making regarding creation and content of the site as well as for maintenance of the site. The person or persons responsible for decision-making, creation of the Web site, and maintenance need not be the same as long as communication is good. The organization must realize that the job is not complete with the creation of the site. Just as the organization is dynamic and the information about the disease is growing, the Web site should be dynamic. If your site provides links to other sites on the Web, the need for maintaining currency is even greater.

Technology-related issues must be tackled next, including access to the Web, equipment needs, and technical expertise. Does your organization have access to the Web? If not, what are your options for acquiring it? Can you afford to pay for it? Is it available through a local free-net? Will someone donate it? A computer workstation must be available at the organization location for access to the Web. It could also be used for creation and maintenance of the site, if desired.

Once you have access to the Web, you will need to determine where you will mount your site. Will your local commercial Internet provider allow you to mount your site with them? Can you piggyback onto another service organization, a municipal server, a local university or health care provider network?

As you meet these technology-related challenges, the need for technological expertise is evident. That expertise may come from the staff, Board of Directors, or volunteers. The level of expertise required will depend on individual circumstances. Many Web building packages now claim that you do not need even a basic knowledge of HTML, although that basic knowledge would certainly be valuable.[2]

Funding is another factor that must be considered. According to a recent *Computerworld* article, the average first-year cost for creation and maintenance of a World-Wide Web site in the corporate arena is listed at $109,000.[3] While having this funding available would be optimal, it is obviously not realistic. The process for a community service organization will more closely match that of the small-business owner who created his Web page through the efforts of volunteers, his young relatives, and by downloading shareware and freeware from the Internet.[4]

Staffing and technology are also funding issues. As discussed earlier, the answer to the "who" question in this process may be greatly dependent on funding. Can you afford to dedicate staff time to this project? Can you afford to hire an outside consultant to provide the design and creation of the site?

In considering technology, we must look at finance as well. Do you have financial resources available to cover the costs of access to the Internet and the necessary equipment? Do you need to purchase software to aid in the creation of the site? What costs will be involved in the maintenance of the site?

Other options may exist to help bear the costs of moving onto the Web. Grants may be available that would provide funding for hardware, software, and/or personnel. Collaborative efforts may help the service organization share the costs of disseminating information via the Web. Perhaps a partnership with a local library committed to AIDS education or consumer health information in general could include a Web project. What other groups has your organization worked with in the past whose purpose would be served by your presence on the Web? Would they be willing to support this project or work collaboratively?

Although, as we can see in the corporate world, moving onto the Web can be a very expensive effort, it does not have to be. Volunteers, collaboration, and creativity can produce respectable results on a shoestring.

After solving the problems or answering the questions related to the mechanics of the Web site creation and maintenance, you are ready to tackle the critical issue, the content of your site. The content of your site should be driven by the answer to an obvious question. What is the purpose of your Web site? In deciding the purpose of your Web site, consider the mission statement for your service organization. A close match between the purpose of your Web site and the organization's mission statement will result in a more focused presentation and a better site.[5]

Your Web site can provide a variety of information related to the organization, its services, and AIDS. Your site's organization may reflect the organization of your services. It may provide general information, educational information, announcements about upcoming events or new services, an invitation for volunteers, and links to other Web sites. One thing to consider as you create links to other sites is the nature of the Web. Sites change or move; some sites become dated because they are not maintained for currency; and new sites appear. If your organization decides to link to other sites, commit to keeping these links current.

Audience should also be considered as you design your site. For whom is the information provided on your Web site intended? Once again, it would be valuable to utilize your mission statement. Is your organization committed to supporting the needs of the gay and lesbian community, all persons with AIDS, and families and friends of persons with AIDS? Are you trying to provide basic AIDS information to educate the general public? Are there other special groups, such as gay and lesbian youth, who also need to be considered in the design of your Web site? What is your organization's link to the local health care providers? The Web's audience in general is broad, but design of your site should be focused on providing a logical, easily-maneuverable, well-organized, visually pleasing site for your primary audience.

When considering the content of your site, your organization must decide if the purpose of the site is solely to provide information about your organization and its services, or also to serve as a resource for other AIDS-related information. In a study of Texas community-based HIV/AIDS service providers, a number of information needs were identified as significant. Consumer education information, drug information, treatment information, health care cost information, and legal information represent categories of those significant information needs.[1] The information you choose or are able to provide or link to will depend, of course, on how you define your Web site's purpose.

The National Library of Medicine's Grateful Med, providing access to various electronic databases including MEDLINE as well as a selection of AIDS-related files, is now available for free access via the Internet. Although these databases represent a wealth of information, they are not very user-friendly for the novice Web explorer. In addition, much of the information retrieved is very technical in nature. Given the diversity of Web users, however, these databases would be a valuable resource to a slice of Internet information seekers.

In considering the content of your Web site, you may wish to begin conservatively. You may wish to start small, gain some experience in the process, and assure that you have the support to maintain the Web site. An important thing to remember in the design phase is the ease with which you can expand the site. If your site design lends itself to the smooth addition of new information and links in the future, your Web presence can be dynamic and keep pace with the changes within your organization and current research on HIV/AIDS.

In perusing the Web, one can see that community-based HIV/AIDS service organizations around the country are in various stages of Web site development. In some of the very large cities, the organizations may have had the financial and human resources that allowed them to move on to the Web early. Although AIDS is misperceived as an urban, gay, white, male disease, in most of America's mid-sized cities the number of AIDS cases is increasing and the organizations are searching for additional ways to educate and inform. Turning to the Web is one option. The Community Health Trust in Louisville, Kentucky, is one of these middle-America cases.

The Community Health Trust was founded in 1984 and is the oldest and largest non-profit AIDS service organization in Kentucky. The Louisville metropolitan area has a population of 991,765[6] and through June 1997 reported a cumulative total of 1142 documented cases of AIDS.[7] This is more than a 100% increase in AIDS cases in a four-year period.[8]

Community Health Trust's mission is to support the health care needs of the gay and lesbian community and all persons with HIV/AIDS, as well as to promote well-being through advocacy, organizational, and educational activities.[9] Community Health Trust offers a number of direct service programs such as the Medicine Assistance Program, delivery of meals to homebound persons with HIV/AIDS, and the Buddy Program, through which trained volunteers provide emotional support, companionship, recreational opportunities, and help with light chores and transportation. The Glade House, a residential program, offers a home for a limited number of persons with HIV/AIDS. Community Health Trust's prevention efforts include provision of safer sex kits to area gay and lesbian bars.[10]

The Community Health Trust publishes a bimonthly newsletter, *Heart Beat*; houses the AIDS Resource Center, a small library of print and non-print materials; and operates the Gay, Lesbian & Bisexual Hotline of

Louisville. Prevention specialists on staff strive to raise public awareness and educate the community about HIV/AIDS.[10]

The Community Health Trust is in the very early stages of moving onto the Web. Dr. Emery Lane, President of the Community Health Trust Board of Directors, recently described the organization's progress in this area.[11] With a staff of thirteen, which includes care coordinators and employees at the residential facility, volunteers will play an important role in the process. Activity on this project temporarily halted when one of the major participants in the planning group, the only staff member involved directly, left the organization. With support from Board members and other volunteers, Community Health Trust intends to regroup and continue with the project. At this stage no firm decisions have been made regarding who will create or maintain the site. However, it appears that they will depend heavily on volunteers.

Some of the technical issues have been resolved. Internet access has been donated by a local Internet provider. The technical expertise necessary for the project should be available from local volunteers. Since Community Health Trust is in the early stages of site development, most of the nuts and bolts technical issues, such as software needed, have not been tackled.

Funding is expectedly a consideration. Application was made for a grant to fund the project; unfortunately, Community Health Trust was not awarded the grant. The Community Health Trust Board of Directors has designated a small amount of dollars for setup and maintenance of the site. This, in conjunction with donation of the Internet access and contribution of effort from volunteers, should allow work on the project to continue.

Perhaps the greatest amount of thought to date has been spent on the content of the site. Tentative plans have been made to provide information about the Community Health Trust and CHT-The Store which donates revenue from the sale of good quality nearly new merchandise. The Web site will provide access to information from the newsletter and will be used to post announcements and notice of upcoming events. Education will be a major focus. Community Health Trust's site will also target reaching gay and lesbian youth who are one of the at-risk populations due to factors such as lack of peer acceptance and risk of suicide.

Community Health Trust committee chairs are being considered as a primary source of materials submitted for the Web site. The Glade House and the newsletter as well as the Education/Prevention Committee and the Client Services Committee will be particularly important.

Louisville's Community Health Trust, like many other AIDS service organizations in mid-sized cities around the country, has just begun the process of creating a presence on the Web. As with any new endeavor, the project will present its challenges. The explosion of the World-Wide Web into today's society has added another dimension to how we acquire information. It is already becoming a mainstream resource. With the continuing need to educate and inform about HIV/AIDS, community-based service organizations are naturally looking at the Web as a means of reaching a broad audience without physical or geographic boundaries.

REFERENCES

1. Huber, Jeffrey T., and Machin, Mary S. "Assessing the Information Needs of Non-Institutionally Affiliated AIDS Service Organizations in Texas." *Bulletin of the Medical Library Association* 83 (April 1995): 240-3.

2. Seymour, Jim. "Building Web-Sites: a Top-Down Job." *PC Magazine* 16 (23 September 1997): 93-4.

3. "Web Site Costs." *Computerworld* 31 (31 March 1997): 69.

4. Coward, Cheryl. "The World-Wide Web on a Shoestring Budget: Putting Your Company Online Doesn't Have to Be Expensive." *Black Enterprise* 27 (June 1997): 58.

5. Cunningham, Jim. "So You Want to Put Your Library on the Web? (How Libraries Should Plan, Implement, Manage World-Wide Web Sites)." *Computers in Libraries* 17 (February 1997): 42-5.

6. U.S. Bureau of the Census.

7. Centers for Disease Control and Prevention. *HIV/AIDS Surveillance Report* 9, no. 1 (midyear 1997): 6.

8. Centers for Disease Control and Prevention. *HIV/AIDS Surveillance Report* 5, no. 2 (July 1993): 4.

9. "Mission Statement." *Heart Beat* 8 (September 1997): 2.

10. Community Health Trust: AIDS and Wellness Services (brochure) Louisville, KY: Community Health Trust.

11. Lane, Emery. Personal interview by author, Louisville, KY, 25 September 1997.

AIDS Service Organizations
and Their Presence on the Internet

Janet A. Ohles

Janette Pierce

SUMMARY. AIDS service organizations have been the driving forces behind providing HIV/AIDS-positive individuals and the public with up-to-date information about the disease, treatment regimens, and prevention measures. It is critical that these organizations develop and maintain Internet sites for the rapid dissemination of information. The Internet offers the capability of providing a communication and publication means to reach a wider audience, reach greater numbers of HIV/AIDS-positive individuals, and reach even those in the most remote areas. This article discusses six major AIDS service organizations (Gay Men's Health Crisis, San Francisco AIDS Foundation, AIDS Project Los Angeles, AIDS Action Committee of Massachusetts, AID Atlanta, and the Howard Brown Clinic) and their presence on the Internet. All six organizations have made a national impact on the provision of HIV/AIDS services, programs, information, and advocacy efforts. *[Article copies available for a fee from The Haworth Document Delivery Service: 1-800-342-9678. E-mail address: getinfo@haworthpressinc.com]*

Janet A. Ohles, MLS (ohles@auhs.edu), is Senior Database Manager, ECRI, Plymouth Meeting, PA. She was formerly the Outreach Librarian at Allegheny Hahnemann Library, Allegheny University of the Health Sciences, Philadelphia, PA. Janette Pierce, MLS (pierce@critpath.org), is Public Service Specialist, AIDS Information Network, Philadelphia, PA.

[Haworth co-indexing entry note]: "AIDS Service Organizations and Their Presence on the Internet." Ohles, Janet A., and Janette Pierce. Co-published simultaneously in *Health Care on the Internet* (The Haworth Press, Inc.) Vol. 2, No. 2/3, 1998, pp. 11-24; and: *HIV/AIDS Internet Information Sources and Resources* (ed: Jeffrey T. Huber) The Haworth Press, Inc., 1998, pp. 11-24; and: *HIV/AIDS Internet Information Sources and Resources* (ed: Jeffrey T. Huber) Harrington Park Press, an imprint of The Haworth Press, Inc., 1998, pp. 11-24. Single or multiple copies of this article are available for a fee from The Haworth Document Delivery Service [1-800-342-9678, 9:00 a.m. - 5:00 p.m. (EST). E-mail address: getinfo@haworthpressinc.com].

BACKGROUND

On June 5, 1981, the Centers for Disease Control *Morbidity and Mortality Weekly Report* recognized the existence of what would become known as the Acquired Immunodeficiency Syndrome (AIDS) in a report on pneumocystis pneumonia.[1] The first AIDS service organizations (ASOs) began as grass-roots, volunteer efforts in response to this new disease–a disease that was met with homophobia from federal, state, and local governments, as well as traditional community and health care channels. Services typically provided by other agencies, such as food, emergency funds, support groups, and education, were offered by AIDS service organizations to fill the needs of a community. In 1982, one year after the recognition of what later became known as HIV, Larry Kramer would write in the second issue of the Gay Men's Health Crisis's (GMHC) newsletter:

> (We have) prepared and distributed 30,000 health recommendation brochures . . . fielded almost five thousand hotline emergency calls . . . created a patient service program that includes crisis intervention counselors . . . individual and group support therapy; a network of buddies to visit and do chores for those who are ill; legal advisers, and financial aid guidance through the complexities of the welfare system . . . set up training seminars for doctors, psychiatrists, psychologists, social workers, health-care-professionals and concerned laymen in all areas of AIDS concerns . . . And all the above services have been provided by volunteers and provided free.[2]

AIDS service organizations' volunteer staff quickly became augmented by professional social service specialists. Most of these agencies became formalized non-profit organizations within a year after their inception.

Major ASOs have developed Internet sites that facilitate the sharing of information between themselves and provide information for HIV and AIDS-positive individuals.[3] HIV/AIDS information is produced at the individual, local, national, and international levels.[4] National and international organizations and agencies have quickly adopted the Internet as a publication and communication tool. It is important that individual and local (e.g., ASO) producers of HIV/AIDS information also utilize the Internet as a publication and communication tool. The Centers for Disease Control and Prevention (CDC) National AIDS Clearinghouse Resource Database (RDIR) of community-based AIDS service organizations had 18,858 records in July 1997. This article discusses six major AIDS service organizations (Gay Men's Health Crisis, San Francisco AIDS Foundation,

AIDS Project Los Angeles, AIDS Action Committee of Massachusetts, AID Atlanta, and the Howard Brown Clinic) and their presence on the Internet. These six organizations represent ASOs that are the subject of numerous published articles and books. They were established as local agencies, but achieved national recognition for their ability to develop publications, databases, and model programs.

INTRODUCTION

Community-based organizations (CBOs) have been defined by various authors. However, the essential elements are the same.

> Organizations that have no formal government attachments either with or beyond local governments and exist totally within the confines of a given community. CBOs are crisis driven and deal with immediate problems.[5] . . . [S]elf directed organizations whose goal is to meet the community's self-defined need. The CBO is located within the community, its leadership comes from the community, and it is staffed by members of the community being observed.[6]

Community-based AIDS services organizations (ASOs) can be defined in the same way. They were started by a community of gay men, were staffed by gay men and their friends, and originally the services provided were for the gay community. Information about locating local service organizations is online in the HIV/AIDS: Putting the Pieces Together Workshop Manual (http://www.auhs.edu/~AIDSinfo).

AIDS service organizations were not the first volunteer or community-based organizations established in the gay community. Pre-AIDS volunteerism in the gay community began in the late 1960s in the areas of civil rights and mental health. By the 1970s, gay and lesbian mental health centers were being established and hot lines specific to gay issues were created. Individual, couple, and group therapy sessions were initiated, focusing on the needs of a gay clientele. The concept of anonymity, so important in the AIDS-affected community, had its start in the needs of closeted gays. Discrimination was a primary fact of gay and lesbian life long before the start of the HIV pandemic. The early belief that AIDS was a gay-related disease resulted in the rapid development of new organizations in the gay community devoted to coping with the impact of the disease. Since the recognition of HIV/AIDS in the early 1980s, AIDS service organizations have been the driving forces behind providing both HIV and AIDS-positive individuals and the public with the most up-to-date information. Major AIDS service organizations have developed Internet sites that provide information and promote advocacy efforts. The Internet offers capabilities to combat the AIDS pandemic through dissemination of information to physicians, community groups, and patients; supplying support groups for patients and families; organizing HIV/AIDS information; providing a means of communication among researchers; and facilitating the research process.[7]

THE IMPACT OF THE INTERNET

The Internet has expanded AIDS service organizations' ability to reach not only a wider audience, but also greater numbers of affected individuals, including those who live in remote or isolated areas. The Internet also allows for an anonymous source of information with a disease that carries with it many social stigmas and prejudices. As noted by Moy et al., "The anonymous inquiry feature of the Internet . . . is a good facilitator of information transfer when addressing sensitive issues involving sexuality and substance abuse."[8]

AIDS service organizations provide reliable information that meets the needs of their clients. Fact sheets posted by Gay Men's Health Crisis (http://www.gmhc.org) present information, such as the HIV antibody test, choosing and working with a physician, and AIDS-related opportunistic infections and cancers. In addition to meeting client needs, GMHC's information on safe sex practices provides educational information for all.

The Internet allows for more timely and pertinent delivery of information to people affected by HIV/AIDS.[9] The AIDS Action Committee and AID Atlanta home pages are updated as information becomes available, and the other ASOs discussed in this article update their pages at least on a monthly basis.[10-11] The San Francisco AIDS Foundation has received feedback that clients visit their Web site to locate treatment information and read their publications.[12] AIDS service organizations serve as the primary source of information for many HIV and AIDS-positive individuals. The Internet allows a means to disseminate information as it becomes available.

Recommendations from a 1993 conference convened by the National Library of Medicine to review HIV/AIDS information resources and services included support for community-based organizations.

> Community-based organizations have been providing services to their communities since the early stages of this epidemic. There is a continuing need to do this and even to expand these services as new

individuals and communities become involved. This is true not only for direct health and social services, but for information services also. Recommendation 5.4 calls on NIH [National Institutes of Health] to help community organizations provide HIV/AIDS prevention and treatment information to patients and their families.[13]

One AIDS service organization, the AIDS Action Committee of Massachusetts, received NLM grant funding to provide electronic access to treatment information through the Treatment Library Internet Project.[10]

While acknowledging the value and power of the Internet as a communication and educational medium, it is important to avoid viewing its arrival as an informational panacea. It would be reprehensible to overlook that many ASO clients are very poor and do not have the requisite computers or computer skills to access electronic information sources.[10,14] Also, as Huber's study noted, there is a need to inform individuals about the availability and applicability of electronic resources.[15] A recurring theme during phone conversations with ASO personnel substantiates Huber's claim that simply making the information available is not enough, since the answers for many questions received by ASO staff over the phone are found on the organization's Web site. Nevertheless, community-based AIDS service organizations are often the first point of contact for an HIV-positive person, and may provide an avenue to reach the client population through electronic information sources.

AIDS SERVICE ORGANIZATIONS' WEB SITES

The founding AIDS service organizations have made major contributions in the areas of HIV/AIDS education, advocacy, and support. The Gay Men's Health Crisis, San Francisco AIDS Foundation, AIDS Project Los Angeles, AIDS Action Committee of Massachusetts, AID Atlanta, and Howard Brown Clinic are all early AIDS service organizations that have had a national impact. Discussed below are the resources these six major community-based AIDS service organizations have available on the Internet.

Gay Men's Health Crisis
(http://www.gmhc.org)

Gay Men's Health Crisis (GMHC), New York, was the first AIDS service organization. GMHC was founded in 1981 by six gay men and incorporated as a not-for-profit organization in January 1982. In one year

GMHC went from a group of six volunteers in a living room to a major social service organization. It remains the largest AIDS service organization in the U.S. The organization's original goal was to provide information and raise money for research.[16] Today, the primary goal is to provide direct care to clients. Services provided include crisis intervention, buddy support, crisis management partners, recreation services, support groups, therapy groups, care partner therapy groups, financial advocacy program, legal services, outreach services, AIDS prevention program, educational publications, and information services.

GMHC's brochures and publications have a worldwide distribution, and provide practical advice on coping with HIV/AIDS. Fact sheets and brochures are linked to easy-to-read information paragraphs throughout its Web pages. As mentioned earlier, the fact sheets address an important need of providing confidential, anonymous educational information. The evaluation research department covers a variety of topics, including evaluation of various GMHC services, specific projects (e.g., House of Latex Project), and surveys. Research project results are reported online and a list of current projects is also given.

Treatment News, a monthly newsletter about treatments for HIV and AIDS, is available full-text at the Web site. Newsletter articles cover drug treatments, along with discussing their pros and cons, and descriptions of therapy outcomes. There are also articles that describe opportunistic infections with appropriate treatment regimens. Additional pieces include interviews with leading research pioneers and explanations of clinical trials. GMHC publications are highly regarded and are permanently linked from many other sites, including CDC National AIDS Clearinghouse (http://www.cdcnac.org/aidslink.html#G&B), Blue Cross/Blue Shield of Massachusetts (http://www.bcbsma.com/hresource/healthlinks.html), MedWeb electronic publications (http://www.gen.emory.edu/medweb/medweb.ejs.html# AIDS and HIV), and ILGA Portugal (http://www.ilga-portugal.org/ingles/news.html).

San Francisco AIDS Foundation
(http://www.sfaf.org)

The second ASO established was the San Francisco AIDS Foundation (SFAF), California, in April 1982. The originators were gay community leaders and physicians who established the organization as an all-volunteer, grass-roots effort. Initial services were a telephone information and referral hot line. The Foundation quickly developed a national reputation for the provision of accurate information, and served as a resource for other ASOs forming across the country. In 1982, SFAF and the San Francisco Department of Public Health established a formal contractual relationship to provide educational services. A year later, a contract was established with the State of California to expand outreach efforts in northern California. During 1984, the focus of providing emergency services changed to a commitment for direct services. By the late 1980s, SFAF had begun responding to the paucity of services provided to minority communities. AIDS hot lines began during this period for both the Latino and Filipino populations. The Foundation offers services in treatment, prevention, education, client services, communication and media, and public policy.[17-18]

The SFAF Internet site was created in August 1996 by volunteers, in response to requests from Foundation supporters and clients. The original purpose of the Web site was to distribute treatment and policy information, as well as provide a source for general information about AIDS.[12] SFAF has several major publications for HIV/AIDS positive individuals, including: the *Bulletin of Experimental Treatments for AIDS* (*BETA*), with both English and Spanish versions; *Early Care for HIV* (first edition); and *Positive News*, a newsletter in English and Spanish with general information on treatment, safe sex, nutrition, and lifestyle issues. *BETA* is online at SFAF back to June 1996, with a link provided to back issues on the AEGIS Web site. Along with the online version of *BETA* at the SFAF Web site, there are permanent links to the journal from Bastyr University's AIDS Research page (http://www.bastyr.edu/research/ARC_resch.html) and from HIV InfoWeb (http://library.jri.org/library/news/beta). The *Outreach* newsletter began publication in 1997, covering general information important to San Francisco HIV-positive individuals, such as financial assistance available for drug regimens, and HIV-specific housing programs. There is extensive hypertext linking both within the Foundation's Web site and to outside documents. For example, an *Outreach* article, "Can HIV be stopped within first 72 hours?" has a link to the original *New England Journal of Medicine* article. Also available is information on the New Prevention Campaign, Compass–the first prevention case management program for gay men. The SFAF Internet site provides valuable information on finding HIV/AIDS resources in San Francisco. It also has effectively provided access to extensive online information and full-text online journals of importance to individuals who are HIV and AIDS-positive throughout the world.

AIDS Project Los Angeles
(http://www.apla.org)

AIDS Project Los Angeles (APLA), California, began as a volunteer project in October 1982 by four individuals who responded to friends becoming ill with the disease, then called GRID (Gay Related Immune Deficiency). The first service was a hot line, established to answer questions, run by twelve volunteers. It became a nonprofit community-based organization in January 1983. Services rapidly expanded throughout the early years. Client services became available when a social worker was hired in 1983. Housing services became available in 1984; a dental clinic was established in 1985; and a food voucher program in 1986.[19] In 1997, this non-profit community-based organization has 230 employees and provides direct service to thousands of people.

The Internet site was begun in late 1995 through the work of volunteers. The move to providing information on the Internet was to keep up with disseminating information through current technology. The original purpose of the Internet site was to provide information about HIV, APLA, and to expand access to its publications. The Web address is made available to clients and there is specific information on the site for clients. Financial and personnel restrictions have precluded the addition of e-mail capabilities, frequent updating, and having more links at the site. APLA's Web site is updated at least once each month.[14]

APLA's Internet presence includes a wealth of information on both the organization and local Los Angeles resources. The site has background information on services, special events, grass-roots efforts, and publications. Services listed range from community education forums available to the general public to case management, assessment, and buddy program services for APLA clients. Housing is a major concern for AIDS service organizations and other organizations dealing with AIDS clients. The Web site has an extensive listing of Los Angeles housing resources from agencies that have roommate listings, to Section 8 housing, to short-term rental assistance, to hospice and palliative facilities.

Several online publications are available. *HIV L.A.* is an extensive listing of Los Angeles services. *Catalyst Newsletter* is a publication for policy change. *Positive Living* has a myriad of general information providing local and national news related to HIV/AIDS, community forums, the APLA movie schedule, benefit information, and general articles. The *Optimist* concentrates on the humanistic side of HIV/AIDS; portraying personal stories from AIDS-positive individuals, advocate volunteers, and APLA employees. Along with giving online access to these journals, APLA's Web site is a model of organizing community information.

AIDS Action Committee of Massachusetts (http://www.aac.org)

AIDS Action Committee (AAC), Boston, Massachusetts, begun in January 1983, is New England's oldest and largest AIDS service organization. The AAC is the largest source of non-governmental funding for AIDS services in Massachusetts, employing 100 professional staff and serving 1,500 men, women, and children. AAC's Youth Only AIDS Line is the nation's first toll-free, statewide AIDS hotline, staffed by teenagers serving teenagers.

AIDS Action Committee's Web site was created pro bono by a design company in May 1995, with AAC volunteers creating HTML documents. A National Library of Medicine grant to provide HIV/AIDS treatment information electronically was the impetus for creating the Internet site. The Web site provides brief descriptions of various services and programs, covering client services, education, public advocacy, support and contributions, and an HIV Treatment Program. E-mail addresses are provided throughout to receive additional information on any program or service. Helpful guidelines to create a resource library for an organization are given on the Web site.

Of particular note is the AIDS Action Treatment Library Internet Project that receives grant funding from the National Library of Medicine and is available on the Web site. The Resource Library gives people with AIDS the most up-to-date information available about treatments. The majority of articles cited are from community-based journals that are not indexed elsewhere. The Resource Library Index gives a subject approach to AIDS treatment information. For example, articles may be found about immune globulins, delavirdine, or integrase inhibitors. The cumulative index focuses on articles about AIDS treatment that will assist people with HIV in choosing medical treatment for themselves, and has a maximum coverage of the last five years. As one would anticipate, an anonymous Web counter on the Resource Library, Index, and Clinical Trials pages shows that they have been accessed thousands of times.[10] Permanent links have been established to AAC Web pages from other Web sites, such as other AIDS service organizations (Provincetown Homepage at http://www.pown.com/newengland/), a family physician practice (http://www.familymed.com/resource.html), virtual libraries (http://library.jri.org/ and http://www.libraries.wayne.edu/dcal/wwwres.html), and Santa Clara County Medical Association (http://www.sccma.org/medinfo/c1.html).

AID Atlanta
(http://www.aidatlanta.org)

AID Atlanta, Georgia, began as a grass-roots effort by the gay community in 1982. It is the nation's first and largest southeastern AIDS service organization. AID Atlanta services include case management, a volunteer "buddy" network, support groups, housing support, and educational programs for the general public and high-risk groups.

The AID Atlanta Internet site was created in 1996 with organizational funds, in response to the general move toward electronic communication. The initial site reflected the annual report, and has been redesigned, presenting information from an organizational perspective. AID Atlanta's Web site is advertised in local magazines, generating requests from individuals on how to volunteer and where to send donations. The site is updated on an ongoing basis under the direction of John Cooper, Systems Manager, and with the assistance of volunteer staff.[11] There are permanent links to its Web pages from the American Psychiatric Association (http://www. psych.org/clin_res/aids_site.html_), the Stop AIDS Project (http://www. stopaids.org/Otherorgs.html), Viaticus Inc. (http://viaticus.net/support.html), and MedWeb (http://www.gen.emory.edu/medweb/medweb.agencies.html# AIDS and HIV).

Treatment breakthroughs in AIDS have resulted in the need for ASOs to offer services addressing issues encountered in living with AIDS, such as financial planning and returning to work. An overwhelming response to a community forum held by AID Atlanta's Department of Education, along with positive evaluations received, led to the formation of the Reconstruction Project. Summaries of the Reconstruction forums are available via the Internet (http://www.aidatlanta.org/reconstr.htm). AID Atlanta has received phone calls from other ASOs about the Project and has provided assistance in answering questions, identifying barriers, and providing expert advice.[3] Reconstruction is a model program for AIDS service organizations who face an increasing need to develop programs for individuals living with AIDS.

Howard Brown Health Center
(http://homepage.interaccess.com/~hbhealth/)

Howard Brown Health Center (HBHC), Chicago, Illinois, was established in 1974 as a clinic to treat sexually transmitted diseases. In the early 1980s, the focus changed in response to the HIV/AIDS pandemic. Services provided on HIV/AIDS include anonymous HIV testing and counseling, primary care medical services, case management and counseling, education, and research.[20] In 1997, research projects included a Multicenter AIDS Cohort Study (MACS), Vaccine Preparedness Study for Sexually Active Men (SAM), and Awareness Intervention for Men (AIM).

Prompted by requests from clients, and through the interest of Diane Goodwin, Media Relations, the Web site was created in 1996 by volunteers. The original, and continuing, purpose of the site is to provide an avenue for community education. The Internet site has increased volunteer support and provides a good presentation of the agency. Both local and national AIDS clients visit the site. Local and national requests for information have been received and answered via e-mail. The Web site has resulted in non-clients requesting information about the Clinic and volunteer opportunities. Lack of funding has restricted the Clinic's ability to expand information on the Internet regarding research activities and educational information, and has prevented the development of interactive capabilities. It is hoped that grant monies will be secured to develop additional Internet resources.[21]

Howard Brown's Web site has general information regarding HIV/AIDS client and outreach services. Summaries of research conducted by HBHC on gay men's health issues, which has given the Center international prominence, are available online. There are three current research initiatives. Funded by the National Institutes of Health, the MultiCenter AIDS Cohort Study (MACS) began in 1983. MACS is the world's largest epidemiological study about sexual practices and how they relate to the transmission of HIV. The HIVNET Vaccine Preparedness Study is a baseline epidemiological and behavioral study. The Awareness Intervention for Men is a series of workshops designed to help gay and bisexual men reduce the impact alcohol and other drugs have on their sexual behavior. Howard Brown is a leader in community-based research concerning the epidemiology, behavior, and clinical aspects of HIV/AIDS.

CONCLUSION

The 1990s is a decade with both great strides and serious disappointments for the HIV/AIDS affected community and the organizations that support them. With the advent of protease inhibitors and cocktail therapies comes the possibility that AIDS-positive individuals will be able to live longer and more productive lives. It has also given the public a false sense that the cure for AIDS is rapidly forthcoming. Regretfully, this belief has translated to lesser monies being donated to AIDS service organizations, while more monies are required to develop additional programs and services for those who are now living with, not dying from, AIDS. The value of the visibility of AIDS service organizations on the Internet and the technology's ability to deliver timely, cost-efficient, anonymous information throughout the world increases in an era of decreasing funds.

The World Health Organization fact sheet HIV/AIDS: The Global Epidemic (http://www.unaids.org/highband/document/epidemio/situat96.html) reported 3.1 million new HIV infections and 22.6 million people living with HIV/AIDS in 1996. Given that these numbers constitute only an estimated one-fifth of all AIDS cases worldwide, the ability for AIDS service organizations to post educational, advocacy, program, and treatment information on the Internet is critical. AIDS service organizations have produced valuable educational programs and services, journals, brochures, and databases to combat the spread of the disease; and to help improve the quality of life and treatment for those affected. The ability to facilitate the sharing of information and replication of model programs is enhanced through this electronic medium. AIDS service organizations' presence on the Internet has made it possible to provide these various resources from one electronic interface. It is especially important, when AIDS service organizations are the primary source presenting medical information in a non-technical, consumer-oriented approach, that the information be presented through the Internet, a medium that has the potential to reach anyone, anywhere in the world.

REFERENCES

1. Anonymous. "Pneumocystis pneumonia–Los Angeles." *MMWR-Morbidity & Mortality Weekly Report* 30 (June 5, 1981): 250-2.

2. Kramer, Larry. *Reports for the Holocaust: The Making of an AIDS Activist.* New York: St. Martin's Press, 1989.

3. King, Mark, Director of Education, AID Atlanta. Telephone conversation with Janet A. Ohles, 31 July 1997.

4. Huber, Jeffrey T., and Machin, Mary S. "Assessing the Information Needs of Non-institutionally Affiliated AIDS Service Organizations in Texas." *Bulletin of the Medical Library Association* 83 (April 1995): 240-3.

5. Lukenbill, W. Bernard. *AIDS and HIV Programs and Services for Libraries.* Colorado: Libraries Unlimited, Inc., 1994.

6. Van Vugt, Johannes P., ed. *AIDS Prevention and Services: Community-based Research.* Connecticut: Bergin & Garvey, 1994.

7. Marlink, Richard, Moderator. "AIDS and the Internet: Networks of Knowledge and Infection (http://www.harvnet.harvard.edu/online/notes/aids.html)." Harvard University Conference on the Internet and Society (May 28-31, 1996).

8. Moy, E.V.; Chu, K.; Frish, I.; Woodward W.; and Wood, J. "AIDS Prevention Programs on the Internet: Interactive Resources in English and Chinese." Abstract No. Th.C. 4558 in *International Conference on AIDS* 11 (July 7-12, 1996): 341.

9. Keil, L., and Hartfield, K. "A Model Electronic HIV/AIDS Information Resource Center." Abstract No. Th.C. 4556 in *International Conference on AIDS* 11 (July 7-12, 1996): 341.

10. Erbland, Peter, Communications Manager, AIDS Action Committee. Telephone conversation with Janet A. Ohles, 28 July 1997.

11. Cooper, John, Systems Manager, AID Atlanta. Telephone conversation with Janet A. Ohles, 29 July 1997.

12. Gordon, Derek, Director of Communications, San Francisco AIDS Foundation. Telephone conversation with Janette Pierce, 1 August 1997.

13. National Institutes of Health (U.S.) Office of AIDS Research. *Information services for HIV/AIDS: recommendations to the NIH: report of a conference co-sponsored by the National Library of Medicine and the NIH Office of AIDS Research, June 28-30, 1993.* Besthesda, MD: National Institutes of Health, 1994.

14. Klosinski, Lee, Director of Education, AIDS Project Los Angeles. Telephone conversation with Janette Pierce, 1 August 1997.

15. Huber, Jeffrey T., and Giuse, Nunzia Bettinsoli, "HIV/AIDS Electronic Information Resources: A Profile of Users." *Bulletin of the Medical Library Association* 84 (October 1996): 579-81.

16. Katoff, L., and Dunne, R. "Supporting People with AIDS: The Gay Men's Health Crisis Model." *Journal of Palliative Care* 4 (December 1988): 88-95.

17. Arno, Peter S. "The Nonprofit Sector's Response to the AIDS Epidemic: Community-based Services in San Francisco." *American Journal of Public Health* 76 (November 1986): 1325-30.

18. Carrillo, Hector, Isabel Auerbach, John Maguire, Carlos Petroni, and Ruth Schwartz. *AIDS Hotline Training Manual*, San Francisco: San Francisco AIDS Foundation, February 1994.

19. Sonsel, George E. "Case Management in a Community-Based AIDS Agency." *Quality Review Bulletin* 15 (January 1989): 31-6.

20. Doherty, John P. "AIDS: One Psychosocial Response." *Quality Review Bulletin* 12 (August 1986): 295-7.

21. Goodwin, Diane, Media Relations, Howard Brown Health Center. Telephone conversation with Janette Pierce, 1 August 1997.

APPENDIX
Interview Questions for ASOs

Organization: ——————————————— Date: ———————————

Contact Person: ———————————— Position: ——————————

1. When was the Internet site created? By volunteers, paid staff, consultant?

2. What prompted the Internet site being built?

3. What was the original purpose of the Internet site?

4. Have any benefits arisen? (i.e., increased funding, increased local support, increased federal support, increased volunteer support, increased client support)

5. Do your clients visit your Internet site?

    a) If yes, for what purposes?

    b) Any feedback?

6. Have you received any feedback that non-clients have visited your site? If yes, what feedback have you received?

    a) Local:

    b) International:

7. Do you know if there are permanent links to your site?

    a) If yes, are these links from other AIDS organizations, other service organizations, federal government, local government, educational, personal Web pages, or corporate Web pages?

8. Did you have grant funded support to build the Internet site or for resources on the Internet site?

9. Is there any information not on your site, due to personnel or financial restrictions, that you would like to have?

10. Do you collect demographics of the people visiting your site? Educational/commercial/governmental/network/personal

11. How frequently do you update the pages?

Building an HIV Internet Network of Community-Based Organizations

Stephanie L. Normann
Donna Rochon

SUMMARY. The primary goal of the Houston AIDS Information Link (HAIL), a consortium of twelve community-based organizations (CBOs) representing national, state, county, city, and public and private non-profit agencies, is to bring current, scientifically accurate information to HIV-infected patients, their health care providers, and the affected community at places where these individuals are already spending time, such as clinics and libraries. HAIL (http://www.NeoSoft.com/~hail/) represents a working example of using the Internet to create a physical and virtual network of HIV community-based organizations, bound together by a common goal. *[Article copies available for a fee from The Haworth Document Delivery Service: 1-800-342-9678. E-mail address: getinfo@haworthpressinc.com]*

PURPOSE

For those of us working in the field of HIV, accessing information about the disease has become second nature. We know where to look for preven-

---

Stephanie L. Normann, MSLS (snormann@utsph.sph.uth.tmc.edu), is Director of the University of Texas Houston School of Public Health Library and Project Director, Houston AIDS Information Link, Houston, TX. Donna Rochon, MA (rita@centerforaids.org), is Editor, Research, Initiative/Treatment, Action (RITA), and doctoral candidate at the University of Texas Houston School of Public Health.

[Haworth co-indexing entry note]: "Building an HIV Internet Network of Community-Based Organizations." Normann, Stephanie L., and Donna Rochon. Co-published simultaneously in *Health Care on the Internet* (The Haworth Press, Inc.) Vol. 2, No. 2/3, 1998, pp. 25-38; and: *HIV/AIDS Internet Information Sources and Resources* (ed: Jeffrey T. Huber) The Haworth Press, Inc., 1998, pp. 25-38; and: *HIV/AIDS Internet Information Sources and Resources* (ed: Jeffrey T. Huber) Harrington Park Press, an imprint of The Haworth Press, Inc., 1998, pp. 25-38. Single or multiple copies of this article are available for a fee from The Haworth Document Delivery Service [1-800-342-9678, 9:00 a.m. - 5:00 p.m. (EST). E-mail address: getinfo@haworthpressinc.com].

tion materials and treatment data, who to call for the provision of social services, and to which physicians to refer patients for optimal care. If we can't find the information ourselves, we know who to call to get it. However, for most of the 239,000 people living with AIDS in the United States, this plethora of information is still inaccessible. In fact, except for a very educated and well-informed minority, most Americans are completely unaware of the existence of this storehouse of information. At best, if they have some idea of the scope of data available, they do not know where or how to access it.

The primary goal of the Houston AIDS Information Link (HAIL), a consortium of twelve community-based organizations (CBOs) representing national, state, county, city, and public and private non-profit agencies, is to bring current, scientifically accurate information to HIV-infected patients, their health care providers, and the affected community at places where these individuals are already spending time, such as clinics and libraries. In 1994, funding from the National Library of Medicine's (NLM) AIDS Community Outreach awards provided the means to initiate this project, and subsequent expansion contracts have both validated its purpose and assured its continuation. Brief descriptions of the consortium members, with the date that they became part of HAIL, are listed in the Appendix.

BACKGROUND

In June 1993, a conference co-sponsored by the National Library of Medicine and the National Institutes of Health Office of AIDS Research provided a forum for clinicians, researchers, the media, allied health workers, patients, and the HIV-affected community to discuss their HIV information needs.[1] It was the members of the HIV-affected community that expressed the greatest frustration in identifying and obtaining up-to-the minute information on HIV/AIDS for themselves and their care givers. NLM responded by offering a competitive Request for Quotation (RFQ) for CBOs and libraries who could meet this demand to provide current, accurate information on HIV/AIDS.

Since the onset of the AIDS epidemic, Houston has ranked in the top ten among U.S. cities with the greatest number of AIDS cases. A multitude of AIDS service organizations (ASOs) have sprung up over the last fifteen years in response to the needs of people with HIV infection in Houston and Harris County, addressing issues of housing shortages, information and referral, education and prevention, and case management. Although the various agencies have been very successful at providing direct client services to people with HIV infection, there had been no attempt to coordinate efforts to retrieve and disseminate prevention and treatment information materials. Leaders of Houston's ASOs convened a meeting to address the issue.

One of the participants in this Houston meeting, the AIDS Education and Training Center's (AETC) Information Manager, had also attended the 1993 NLM AIDS Conference. He recognized that the information needs of our community were a reflection of the national pattern and generously offered to share his expertise in identifying the available HIV/AIDS online resources. The Executive Director of Houston Clinical Research Network (HCRN) volunteered to increase Montrose Clinic's role as a resource for patients to obtain HIV/AIDS information. The AIDS Project Director of Harris County Hospital District (HCHD) suggested that an education/training site be located in the PWA Coalition office at Thomas Street Clinic. Although the University of Texas-Houston School of Public Health's (UTSPHL) extensive collection of HIV/AIDS materials had been available on site to the Houston community, what was needed was electronic access from clinics and public libraries within the HIV community.

Each of the organizations participating in the initial proposal had tried at varying levels to satisfy the information needs of its primary target group, but with little overall success. Further, the restricted financial assets of each agency prevented the acquisition of equipment necessary for accessing the Internet. The NLM's RFQ for AIDS Information Outreach Projects provided a vehicle for realizing the consortium's goal. UTSPHL prepared and submitted a proposal on behalf of the CBOs, with the objective of providing access to a wide variety of electronic HIV/AIDS resources through Gopher servers, library catalogs, and bulletin board systems from the specified CBO locations. The purchase and installation of state-of-the-art PCs, equipped with CD-ROM drives, modems, and peripheral printers, were specified. The limited budget ($25,000) outlined in the proposal required significant in-kind commitments, primarily in terms of staff effort, from all of the CBOs involved. Also, the proposal highlighted the importance of constructive interaction between these diverse organizations with similarly focused needs.

ORGANIZING HAIL

Early on, the group established its own name and designed a logo. This not only gave the group a unique identity, but also allowed the efforts of the organizations and individuals at each participating site to reflect on the group as a whole rather than on any specific unit. A steering committee composed of at least one key member from each of the participating agencies was formed. It is responsible for setting its own agenda and priorities. The members decided that the committee would meet on a monthly basis–a pattern to which it continues to adhere. It would address such issues as publicizing the project, training staff, and monitoring the use of end users' responses to various electronic resources. The representatives of the participating organizations have changed over time, generally moving from senior management to individuals responsible for program development and implementation. The quality, depth, and variety of skills that are represented on the steering committee are illustrative of the high esteem that the organizations hold for the program.

The next issue to be tackled was computer equipment and Internet connectivity for each site. UTSPHL and the AETC had full Internet access; Houston Public Library (HPL) had been designated as the city's Internet node, but did not yet have any applications mounted; HCHD and HCRN relied solely on modem connections for dial access catalogs and databases. The user-friendly software Grateful Med, used to access the NLM databases AIDSLINE, AIDSDRUGS and AIDSTRIALS, was ordered and installed on the HAIL PCs at the participating sites. A subscription for the CD-ROM version of AIDSLINE was also entered for each agency. Resources from the Centers for Disease Control and Prevention (CDC), such as NAC Online, and informative BBS's were introduced to the organizations. Training sessions for staff of the participating CBOs were held at the National Network Library of Medicine's South Central Regional office and at UTSPHL.

A major achievement, Internet connections for each CBO, was accomplished in the fall of 1995 through the generosity of a Houston-based Internet service provider, NeoSoft, which offered the project ongoing free Internet access and space on their Web server. This allowed us to mount our HAIL Web page (http://www.neosoft.com/~hail) with its primary link to the AETC's HIV/AIDS Internet resource page. As each participating CBO mounted their own Web page, reciprocal links were created with HAIL. When HPL linked to the HAIL Web page and Internet resources, it effectively gave all of Houston access to Internet HIV/AIDS information resources.

HAIL IN ACTION

From its inception, consortium members displayed synergistic rapport and great enthusiasm for this undertaking. HAIL was announced to the public with a "Media Day" event, held at the Montrose Public Library branch; executive directors of the participating organizations outlined the advantages of the project for their clients. Because the leaders of participating CBOs were anxious to provide their staff and clients with Internet access for the retrieval of HIV/AIDS information, they were willing to give time and resources to ensure the project's success. As a result, HAIL was able to build efficient, supportive relationships between the participating sites, which have proved beneficial for the entire Houston HIV/AIDS infected and affected community.

Establishing HAIL connections at branches of the Houston Public Library and more specialized libraries has greatly increased the availability of health information to the general public. The library buildings are open and staffed with information professionals for extended hours in convenient neighborhood locations. From the curious consumer of medical trivia to the sophisticated investigator, online research topics are supported by a body of literature in many formats; the HAIL workstation is an admirable complement to more traditional library materials.

Although library services staff can assist clients with the technical aspects of Internet searching, they are not permitted to interpret HIV or medical information for clients. This situation highlighted the demand for more understandable and language-specific materials. As a result, HAIL set about identifying CBOs serving culturally diverse populations to address these needs. With subsequent NLM awards, HAIL was able to include two new agencies in 1995: an AIDS advocacy organization, The Center for AIDS (CFA), and Amigos Volunteers in Education and Services, Inc. (AVES), an AIDS service organization whose mission focuses on the Hispanic community. Including AVES not only improved our ability to reach the Spanish-speaking population, but also facilitated Spanish language translations of newsletters and brochures. In 1996, HAIL was able to bring in two agencies that reach truly underserved populations: WAM Foundation, Inc. (WAM), which assists the African American community, and the Harris County Sheriff's Department, Medical Division (HCSD), which serves the incarcerated population of this metropolitan area.

Prior to the HAIL project, none of the ASOs were linked to the Internet. Now, members frequently assist each other with the configuration and mounting of Web pages and linkages to the HAIL page. The AETC has helped several members develop their organization's Web pages, and in turn, has become better acquainted with other Houston-based AIDS organizations through regular monthly HAIL meetings and HAIL training programs. Case managers and educators at AVES received training on how to use the workstation and can now navigate the Internet in search of data, making the Web one of the agency's most valuable tools. They have also begun publishing a bilingual Web page that has allowed the Hispanic community to access information in Spanish. The steering committee members of the HCSD Medical Division who attended a HAIL-sponsored HTML class have recently developed a most attractive and informative Web site on which they are now mounting the bi-monthly newsletter that they create and distribute to the female prisoners. The material for their newsletter, HERR: About Her Health!, is also gleaned from Web sources.

The CFA, which has recently created its own impressive home page with extended links, is also in the process of making its newsletter, RITA!: Research, Initiative, Treatment, Action, available on a national basis. Through their association with HAIL, the publications of both HCSD and the CFA have been added to the CDC's National AIDS Clearinghouse of HIV/AIDS information, a major coup for local organizations. The HAIL workstation is a major feature of CFA's newly opened storefront information center, which offers easy Internet accessibility to all community members. It allows staff, volunteers, and the average patient direct access to a large body of HIV/AIDS science. Thus, this project has accentuated the positive impact that the Internet can have on providing information to members of its community.

The coalition-building spirit within the HAIL consortium may be one of its most impressive outcomes. For participating members, experiences related to HAIL are shared in a meaningful learning environment in which referrals are encouraged between CBOs. In this way, clients can be directed to the agency that best serves their information and service needs. The HAIL workstation has also been used in discharge planning, providing important information for the client moving from one geographic area to another–sometimes across the state, sometimes to another country. Further, the Montrose Clinic's Deaf Program staff has been training clients on computer access so that a much larger portion of the Deaf/Hard of Hearing population can be reached. The HAIL PC at Montrose Clinic has allowed these clients, who rely on visual signs, tools, or gestures for communicating effectively, to gather information on HIV in a self-paced, efficient manner. The Education Division of Montrose Clinic has also included their impressive "Next Step" program on their Web page, thereby allowing access to this information by HIV-infected individuals and health educators in their client group both locally and across the state.

The fastest-growing risk group for HIV-infection in the U.S. is injection drug users and their sex and needle-sharing partners. This population is predominantly young, sexually active, and disproportionately African-American and Latino. As such, the second wave of the HIV epidemic has had a greater impact on minority populations. A 1992 study documented a low level of knowledge regarding HIV/AIDS among minority respondents and this holds true today. Based on feedback that both WAM and AVES staff have received during HIV educational presentations, it is imperative to raise the level of knowledge on HIV/AIDS in the minority community, for whom information materials sensitive to their culture and language are scarce, in order to effect behavior change and slow the increase of AIDS cases in these communities.

With the resources allotted through HAIL, staff and clients of AVES and WAM have access to vital HIV information. Initially, there was very little client use of the HAIL workstation at AVES because the monolingual Spanish-speaking clients were intimidated by a computer. To break this barrier, some case managers began to incorporate Internet training for clients at the time of intake, with the goal of familiarizing clients with the workstation so that they would be able to access health care and treatment information on their own. Staff at AVES are finding that more and more clients have become curious about the Internet and are no longer afraid to give it a try; those clients who periodically use the HAIL workstation report that it is "fun and fascinating." WAM staff use Internet services to receive the CDC AIDS Daily Summary and MMWRs, and to exchange ideas and opinions with other individuals around the globe, thus arming WAM's educators with the necessary tools to combat the myths and misinformation surrounding HIV and AIDS.

HAIL and its steering committee are now reflective of the greater Houston/Harris County population. The experience and focus of the committee members within their individual organizations, as health educators, case managers, patient advocates, treatment information specialists, clinical outreach workers, public librarians, and information managers, suggest the array of services required for the HIV community. The patient advocate office of Thomas Street Clinic feels that HAIL is vital to the clinic's efforts to enhance knowledge and independence of those affected by HIV in Harris County, Texas. Patients with limited mobility can get online answers to their questions in conjunction with their doctor or pharmacy visits, eliminating additional trips to computer sites where free or low-cost transportation is not provided. The majority of Montrose Clinic's clients do not have off-site access to electronic information resources that would help them manage their illness, but the Clinic staff can pass on answers to questions about medications and treatments found from Internet sites. Updated information on HIV is not always readily available to the medical staff at HCSD, but facts about HIV testing, hepatitis, gynecological infections, etc., can be gathered quickly online. The HCSD health educators have even been able to find the answer to some little-known disease or problem and relay this information to the medical staff. STD prevention information can be given directly to their clients. Because there were no library facilities, it was difficult to research the health care needs of clients at the jail. The HAIL workstation has been an invaluable tool in terms of convenience and ease of use.

After four years of operation, the scope and accomplishments of HAIL are reaching beyond the Houston area. Because the coordinating center of the AETC is located at UTSPH in Houston, its staff has consistently participated in numerous HIV/AIDS organization events at both the management and support levels. The UTSPH Library Director and the AETC Information Manager exhibit the HAIL project, highlighting the members' services and demonstrating Internet search software and HIV/AIDS resources (i.e., Internet Grateful Med, AEGIS, CDC, HIV InSite, etc.) at most HIV-related conferences and seminars that take place throughout Texas. Our members take pride in having the HAIL exhibit next to theirs at the annual HIV/AIDS Workshops of the Texas Department of Health and Dr. Adam Rios' Houston AIDS Conferences. Requests for the HAIL exhibit, demonstrations, and workshops have been accepted at community and civic organizations (the Houston Mayor's Neighborhood Connections Conference, National Association of County and City Health Organizations), national, regional and state library association meetings (Medical Library Association, Kansas City; MLA South Central Chapter, Galveston; Texas Library Association, Houston); and physicians' and allied health organizations (Texas Medical Association's Technology Conference, Austin; Area Health Education Centers National Conference, San Antonio). HAIL presented a poster session at the XIth International AIDS Conference in Vancouver in July of 1996[2] and also participated in the Health Resources and Services Administration's Resource Fair. The CFA will highlight the benefits of HAIL during their workshop presentation at this year's National AIDS Treatment Advocates Forum in San Diego in November. ASO representatives from the project periodically speak to community groups about the significance of up-to-date HIV/AIDS medical information, outlining the ways in which HAIL opens up the previously inaccessible world of the Internet to traditionally underserved populations.

EXPANSION OF PROJECT

HAIL's impact on the Houston HIV community has been significant. Not only has it brought previously unattainable technical information to HIV-positive individuals, their care givers, and the affected community at locations convenient for these individuals, but it has been enthusiastically adopted by a cross-section of AIDS service providers and the targeted audience. Our presence on the Web has elicited interest from other CBOs who are actively soliciting to become members of HAIL.

The four consecutive NLM AIDS Information Outreach Project awards have allowed HAIL to broaden and enhance its program, giving it a better balance of participating organizations and offering continued recognition that there are more steps to be taken in connecting Internet end users to the wealth of available HIV/AIDS information. The organizations that comprise HAIL are reflective of the diversity of the HIV-affected community, and it is anticipated that the Galveston-based agencies, being brought into HAIL this year, will extend our impact across the state. The University of Texas Medical Branch at Galveston (UTMB) is the site of the AIDS Care Clinical Research Program (ACCRP) with its NIH-sponsored AIDS Clinical Trials Unit (ACTU). The majority of the HIV-infected individuals treated at the ACTU come from the counties in the surrounding area, but others come from counties located throughout Texas. HAIL workstations will soon be installed in the patients' waiting room at the ACTU and at a Galveston-based AIDS service organization, the AIDS Coalition of Coastal Texas, Inc. (ACCT). UTMB's Moody Medical Library staff will conduct online training sessions for the staff and volunteers of these new HAIL member agencies.

CURRENT FOCUS AND FUTURE PLANS

The HAIL consortium and its agencies reflect the evolving AIDS epidemic: emphasizing the role of information access at clinical treatment sites where there is an opportunity for the HIV-infected to communicate with knowledgeable treatment advocates; increasing training and continuing education in electronic and Internet access for the staff and volunteers of the various CBOs so that new staff have a means of adapting to the sophisticated HAIL resources; presenting exhibits, demonstrations, and workshops using HAIL's linked HIV/AIDS Internet resources at local and regional meetings; endorsing the use of the HAIL home page and linked home pages to describe and promote the mission, programs, and services of the participating agencies; and using appropriate media sources to publicize HAIL–its programs and its accomplishments.

Above all, one must take into consideration that the management of HIV disease changes daily. With the advent of the protease inhibitors and their demanding treatment regimens, people are living longer, healthier lives that were previously impossible. HIV-infected individuals are taking an active interest in managing their own health care and are realizing the benefits of becoming personally involved with their physicians in designing treatment programs that provide maximum benefit with minimum side effects. To work in partnership with their care givers, patients need to educate themselves, and often their doctors, with the most up-to-date information about new developments in HIV/AIDS treatment and research.

The project's success and expansion clearly show that it is meaningful to the target population, and changes in project emphasis show how it is reacting to the evolving face of the HIV epidemic. So many ASOs are struggling to learn the current technologies in order to improve the caliber of direct client services. The productive interactions fostered by HAIL, with the initiative and support of the participating members, will enable the average patient or concerned individual to maneuver through the labyrinth of HIV/AIDS scientific and medical data in a way that will furnish them with vital HIV/AIDS treatment information and will enhance the quality of their lives.

CONCLUSION

It seems appropriate to posit the factors that are contributing to the success of the Houston AIDS Information Link consortium:

- HAIL is a consortium comprised of community-based organizations which represent all of the constituencies in the geographic area. Some of our members focus on special population groups while others serve the members of the entire Houston/Harris County area. This makes steering committee meetings an ideal networking base.
- The consortium is supportive of the mission of each agency and respects their autonomy and independence. The contract manager is a neutral partner.
- Our members are appreciative of the role of HAIL as a facilitator. Planning and decision making are participatory. All HAIL partners have an equal voice in these functions, and it is important that everyone is heard.
- Our monthly steering committee meetings provide the forum for HAIL representatives to share their expertise and discuss their programs with each other. The open communication between all of the agencies promotes client referral.
- We demonstrate this cooperative, interactive spirit on the consortium home page by prominently displaying the hypertext links to all of our partners' Web pages immediately following the HAIL logo.

- The basis of HAIL is functional, not status-oriented. The education and staff training stimulate the outreach programs of our member agencies. The inclusion of both public and health sciences libraries in community-based organization activities is encouraged. Who is better equipped to promote community health information?
- We emphasize the tools of communication, such as Internet access, in providing the means of obtaining current, scientifically accurate HIV/AIDS information, thereby helping the agencies more fully and efficiently meet their goals and objectives. By assisting the organizations in this manner we are helping each ASO recognize that, indeed, they are each making a difference.

REFERENCES

1. National Library of Medicine. *Information Services for HIV/AIDS: Recommendations to the NIH: Report of a Conference Cosponsored by the National Library of Medicine and the NIH Office of AIDS Research, June 28-30, 1993.* Bethesda, MD, National Institutes of Health, 1994.

2. Normann, S.L., and Meyer, J. "AIDS Community Outreach: Houston AIDS Information Link (Hail) Bridges The HIV/AIDS Information Gap." Abstract no. We.D.3779. *International Conference on AIDS* 11 no., 2 (July 7-12, 1996): 188.

APPENDIX
Houston AIDS Information Link (HAIL)
Participating HAIL Organizations, 1994-1997

- The University of Texas Houston School of Public Health Library (UTSPH), 1994, is the lead agency for the project, with management responsibilities for the NLM AIDS project awards. UTSPH is a unit of UT Houston Health Science center and one of six libraries constituting the Texas Health Sciences Library Consortium. It shares a fully integrated online catalog with a WWW interface and significant wide-area search capabilities (i.e., MEDLINE, CANCERLINE, PsychINFO, etc.). UTSPH Library's involvement in the collection and management of HIV/AIDS information was stimulated by its collaboration with the AIDS Education and Training Center for Texas and Oklahoma from the 1988 inception of this federally funded project centered at the UTSPH.
- AIDS Coalition of Coastal Texas, Inc., Galveston (ACCT), 1997. Established in 1985, ACCT serves hundreds of people with HIV in the Galveston, Brazoria, and Matagorda area. The coalition provides

no-cost psychosocial, educational, financial, and nutritional services; and endeavors to identify unmet needs and develop client-oriented programs to meet those needs. Some of its innovative services include washer/dryer facilities, a client resources room, an AIDS information library and home and hospital visits. The agency also works with physicians and other care givers who treat their clients.

- AIDS Education and Training Center for Texas and Oklahoma (AETC), 1994. Established in 1988, the AETC is federally funded to provide education and clinical training to health care professionals regarding current understanding of HIV infection and the treatment of AIDS. In addition, it operates a toll-free telephone AIDS information service, called the AIDS Helpline for Health Professionals, which receives more than 100 calls per month.
- AIDS Foundation Houston, Inc. (AFH), 1997. Founded in 1982, AFH was the first CBO in Houston dedicated solely to HIV issues. It offers HIV/AIDS education, social services, and health care to the infected and affected community on a widespread basis through an extensive volunteer program. Ongoing activities include an HIV hotline, Stone Soup (food pantry), A Friendly Haven (housing for infected women and their children), buddy and hospital teams, Camp Hope (camp for HIV-positive children), and a proposed teen hotline.
- Amigos Volunteers in Education and Services, Inc. (AVES), 1995. AVES provides education, outreach, and direct client services to Hispanics at risk or infected with HIV/AIDS in a linguistically, culturally sensitive manner. Clients receive services such as case management, psychotherapy, support groups, an ethnic-specific food pantry, financial assistance, and wellness activities. They also assist other service providers in translating their written material from English to Spanish and in sponsoring educational programs for the HIV-affected Hispanic community.
- The Center for AIDS: Hope and Remembrance Project (CFA), 1995. Housed in a unique storefront setting, this information and advocacy organization is dedicated to bringing news about the latest medical advances in HIV/AIDS to members of the Houston infected and affected community. It publishes a newsletter, RITA! (Research Initiative/Treatment Action!), sends a weekly fax newsletter that alerts more than 200 health care professionals to late-breaking stories in HIV/AIDS research and treatment, and hosts community forums on a variety of treatment-related topics.
- Harris County Hospital District's Thomas Street Clinic (HCHD), 1994. Thomas Street Clinic, the first publicly-funded HIV outpatient

clinic established in the United States, continues to be the largest provider of HIV medical, pharmaceutical, and psychosocial outpatient services in the state of Texas. The clinic offers a multi-disciplinary system of health care delivery for medically and financially disenfranchised Harris County residents who are HIV-infected.

- Harris County Sheriff's Department (HCSD), Medical Division, 1996. HCSD has an average yearly population of 100,000 and a seropositivity rate of 5.6%. The HIV Department, a separate program within the Medical Division, not only provides counseling and testing for HIV to the general incarcerated population, but also operates a full-time HIV clinic, with case management and support group sessions for positive inmates. Workshops on HIV/AIDS and sexually transmitted diseases (STDs) are held daily at the jail.
- Houston Public Library (HPL), Central and Montrose Branches, 1994. HPL offers a broadly defined program of informational and educational opportunities for Houstonians of all ages and educational, cultural, and economic backgrounds. The library system, serving a diverse population of 1.7 million people, has an online public catalog system with dial-in access and recently implemented Internet access. The Central facility has completed installation of more than 70 PC's with full graphics capability within their building and more than 350 PC's system-wide.
- Montrose Clinic's Houston Clinical Research Network (HCRN) and Community Services Division, 1994 and 1997 respectively. With a full array of ancillary clinical, social, and educational programs, the clinic serves people with HIV in the Houston area by identifying unmet patient needs and developing client-oriented programs, such as low or no-cost health care and research services. HCRN, the research division of Montrose Clinic, is dedicated to providing access to community-based clinical trials on experimental therapies for HIV/AIDS. The Community Services Division has an extensive deaf/hard-of-hearing outreach program targeting this underserved population. It also offers Next Step, a health education and wellness program for newly diagnosed individuals.
- The University of Texas Medical Branch at Galveston, AIDS Care and Clinical Research Program (UTMB-ACCRP), 1997. The oldest medical school established in Texas, UTMB has provided care to indigent patients throughout the state for over 100 years. The ACCRP is the HIV treatment arm of the Infectious Diseases Division of UTMB and serves as an AIDS Clinical Trial Unit site. Since 1982, UTMB's HIV Clinic has provided care to more than 5,000 individu-

als with HIV infection; approximately half of the patients are inmates in the Texas Department of Corrections. Patients may travel to UTMB from all 254 counties in Texas.

- WAM Foundation, Inc. (WAM), 1996. Established in June 1989, WAM is a minority CBO offering an array of educational, social support, and case management services to a predominantly African American clientele. Their mission includes the provision of culturally sensitive, language-specific prevention information on HIV/AIDS and quality support services targeted to the HIV-infected African American community. WAM provides services to 700 unduplicated HIV-infected individuals each year and their educational component educates approximately 400 previously unreached individuals annually.

Web Sites as Weapons in the War on HIV: Education and Prevention Geared to the New At-Risk Populations

Gerald (Jerry) Perry

SUMMARY. The demographic profile of HIV infection has evolved just as the science of anti-HIV therapy has progressed. Rates of new infection are disproportionately rising within minority groups. Continued ignorance about HIV continues to take a grisly toll. The World-Wide Web, as a democratic information distribution system allowing privacy at point of access, may be the best medium available to improve prevention initiatives. This article evaluates a select list of HIV/AIDS education and prevention Web sites geared to the new at-risk populations. An Appendix provides an additional list of general interest HIV/AIDS education and prevention Web sites. *[Article copies available for a fee from The Haworth Document Delivery Service: 1-800-342-9678. E-mail address: getinfo@haworthpressinc.com]*

In the Introduction to the first edition of the *AIDS Information Sourcebook*, I wrote, "Compassion empowered with knowledge can achieve miracles. Short of a miracle, education is the only way to prevent the

Gerald (Jerry) Perry (jperry@ahsl.arizona.edu) is Head of Information Services at the Arizona Health Sciences Library, University of Arizona, Tucson, AZ. Jerry is active in the Medical Library Association (MLA) and was co-convener of the Lesbian/Gay/Bisexual Health Sciences Librarians Special Interest Group of MLA.

[Haworth co-indexing entry note]: "Web Sites as Weapons in the War on HIV: Education and Prevention Geared to the New At-Risk Populations." Perry, Gerald (Jerry). Co-published simultaneously in *Health Care on the Internet* (The Haworth Press, Inc.) Vol. 2, No. 2/3, 1998, pp. 39-51; and: *HIV/AIDS Internet Information Sources and Resources* (ed: Jeffrey T. Huber) The Haworth Press, Inc., 1998, pp. 39-51; and: *HIV/AIDS Internet Information Sources and Resources* (ed: Jeffrey T. Huber) Harrington Park Press, an imprint of The Haworth Press, Inc., 1998, pp. 39-51. Single or multiple copies of this article are available for a fee from The Haworth Document Delivery Service [1-800-342-9678, 9:00 a.m. - 5:00 p.m. (EST). E-mail address: getinfo@haworthpressinc.com].

39

spread of AIDS and save lives."[1] Those sentiments still ring true a decade later.

What has changed, however, is the prognosis for people who contract the AIDS-causing virus, the Human Immunodeficiency Virus or HIV, and the demographic profile of those who are becoming newly infected. Drug "cocktails" of recently developed protease inhibitors, combined with zidovudine (AZT) or similar anti-retroviral drugs, are dramatically increasing the life spans, and postponing the onset of opportunistic diseases, in individuals infected with HIV. People long without hope are experiencing what's now commonly referred to as the Lazarus effect; having given up expecting to live, these people now find themselves with hope for a future. But as hope-inspiring as these new therapies may be, infection with HIV and subsequent contraction of AIDS is still, for most, a fatal disease. From what we know to date, the cocktails do not cure, they delay.

While the science of HIV/AIDS therapy has progressed and evolved, from anti-viral therapies to drugs that block the replication of the virus and decrease overall "viral load" in the infected host, so has the demographic profile of HIV infection changed. Originally a disease focused in the white gay male, intravenous substance user, Haitian and hemophiliac populations, HIV infection is increasingly becoming a disease of poverty, with incidences of new infection occurring most frequently, and alarmingly, in young gay men of color, African-Americans, Latinos and Latinas, Native Americans, and women of color.[2]

Popular science author Wayne Biddle, writing in the introduction to his book, *A Field Guide to Germs*, a delightful if morbid survey of the most common infectious and communicable diseases to have afflicted mankind over the centuries, states, "The triumph of scientific medicine in this century has been to push calamitous encounters with pathogenic organisms away from bourgeois life. Deadly or disabling illness of this sort is more closely associated than ever with poverty, malnutrition, unconventional lifestyles . . . , sexual promiscuity . . . , substance abuse, and other facts of human nature that tend to be ignored by polite circles whenever possible."[3] Ignorance, in fact, is continuing to take its toll, as rising rates of new infection testify, despite well over a decade's worth of HIV/AIDS prevention and education initiatives.

And so the challenge remains. How to prevent new infections with HIV? How to inform and educate the new at-risk populations? How to effect changes in behaviors that stick despite the inclinations of human nature?

In 1986 and 1987, while compiling the lengthy lists of educational and prevention-related materials to include in the *AIDS Information Source-*

book, my co-editor and I did not have the option of including resources on the Internet. It was in existence, but at that time served a fairly select community: science researchers affiliated with the government, defense contractors and academic elites. A lot has changed, and the Internet is now firmly established as one of the most universal information distribution systems known to mankind. With sufficient economic resources, anyone can publish on the Internet, and anyone can retrieve that information. For those without the necessary financial resources, public libraries have begun to make information on the Internet widely accessible. The growth and popularity of the World-Wide Web as a means to graphically display and assist with easy assimilation of information attests to the public's acceptance of, and excitement with, this new medium.

The World-Wide Web, with its paradoxical features as a truly democratic information distribution system while allowing privacy at point of access, may be the best medium available to HIV/AIDS prevention educators. Audience-specific prevention messages can be loaded on servers and accessed remotely by Web surfers located anywhere in the world, working from private workstations in home, office, clinic, and bedroom.

Reaching people "where they live" in language they understand about an issue that can be a matter of life and death is the goal of many HIV/AIDS educators presently active on the net. Web-based services such as The Body, the Safer Sex Page, STOP AIDS Project, and the Centers for Disease Control and Prevention's (CDC) National AIDS Clearinghouse Web sites, among many, are excellent examples of general Internet-based resources available to the HIV/AIDS educator, information professional, and curious lay person seeking information. See this article's Appendix for a list of some of the best currently-available general interest HIV/AIDS education and prevention Internet sites.

Web page search engines, including new meta-search services such as Dogpile (http://www.dogpile.com), now allow Web surfers to utilize multiple search services at a single time, locating long lists of Web resources, ranked by apparent relevancy based on the search string entered by the searcher. Search engines, including the latest meta-engines, are a crucial tool, but even with the best of them there continues to be the problem of specificity and ultimate relevance to the question at hand. Certainly, progress will continue in the perfection of these tools.

In lieu of the "perfect search machine" arriving, this article offers a select and evaluated list of HIV/AIDS education and prevention Web sites geared to the new at-risk populations. Targeted groups include adolescents, African-Americans, Asians, Latinos and Latinas, Native Americans/First Peoples, women, and individuals living in rural communities.

As with any review of Web-based Internet resources, everything and anything listed here is subject to change. Web pages relocate at what seems like whim. Uniform resource locators (URLs) change without notice. Pages are constantly under construction; what was true about a Web page yesterday is false tomorrow. Some pages are allowed to become dated and quickly irrelevant, especially as they concern a rapidly evolving knowledge base such as what we know and are learning daily about HIV/AIDS. Given these caveats, the following sites offered unique value in the ongoing campaign against epidemic HIV/AIDS infection as of September 1997. As you use these and other Web sites, remember to apply your critical evaluation skills and verify accuracy, timeliness, authenticity, and relevancy.

In his *A Field Guide to Germs* entry on HIV, Biddle states, "Unlike the Black Plague, which tended to come and go, AIDS may belong in the same realm as cancer or heart disease. That is, for some but not all populations it is an endemic killer. Or, it may be similar to tuberculosis in the last century. It differs from all three points of comparison because it is totally preventable, communicable but not contagious."[3] The task at hand for AIDS-aware health information professionals, concerned activists, health care professionals, and impacted individuals is to use knowledge and information, increasingly culled from the Internet, to prevent HIV from establishing itself as an endemic disease in America's increasingly Third World-like urban centers and among those marginalized by poverty and the long-term effects of racism and bigotry. We still need a miracle but, short of that, education will have to do.

ADOLESCENTS

CAPS Hot Topic: Adolescents
(http://www.epibiostat.ucsf.edu/capsweb/hotteens.html)

According to researchers at the National Cancer Institute, one in four new HIV infections in the United States occurs in a person under the age of 22.[4] Minority teens are particularly at risk. For youths aged 13 to 19, African-American females accounted for 73% of new infections in 1993, according to the Centers for Disease Control and Prevention.[5] With HIV having such a long latency period, in some instances more than 10 years, many of these adolescents were probably sexually exposed at alarmingly young ages. Certainly one of the saddest tragedies of the HIV/AIDS pandemic is our ongoing failure to prevent these infections. And yet HIV/AIDS educators continue to face a conflicted public, wanting to protect teens but unsure whether giving them information about sexuality might in fact promote sex behaviors. All this, despite data testifying that teenagers are having sex earlier than ever and often with multiple partners, a recognized HIV risk behavior.[6] Fundamental fear of teen sexuality seems to drive this willful ignorance. There are, however, a fearless few on the Internet promoting teen HIV education.

The Center for AIDS Prevention Studies at the University of California, San Francisco, has developed an especially exciting Web-based resource geared to adolescents and HIV/AIDS. Their "CAPS Hot Topic: Adolescents" site starts with a "Chat Line-You Tell Us" service where visitors can register their opinions on a current, often controversial, topic and read what others have written. Most recently, the question was, "Should we teach abstinence only in schools?"

Following are links to a number of featured regional adolescent outreach program Web sites, including: the Healthy Oakland Teens Project, which focuses on the use of teen peer role models advocating for risk reduction; the MPowerment program targeting young gay men and operating out of Eugene, Oregon; and the Youth Action Project based in Southern Alameda County, California, which provides services to gang-affiliated teens, runaways, and other high-risk kids. Other CAPS highlights include bilingual Spanish and English "Fact Sheets," opinion essays, and a bibliography of articles about HIV/AIDS and adolescents.

The CAPS initiative sponsors original research regarding HIV/AIDS prevention, and this site features a listing of full-text summaries about several projects oriented to adolescents, including "Effects of School HIV Prevention: A Secondary Analysis," "Hispanic Parent-Child Interactions Regarding Sexuality," and "Marketing HIV Prevention and Testing to Today's Youth."

While not as frank a resource as the Chicago-based Coalition for Positive Sexuality's "Just Say Yes" Web site (http://www.positive.org/Home/index/html) or as moving as the "Living With AIDS: HIV-Positive Teens Tell their Stories" site (http://desires.com/1.4/Sex/Docs/aids.html), the "CAPS Hot Topic: Adolescents" Web page is an important, continually updated, and authoritative resource that addresses head-on the divisive issue of teen sexuality in an era of AIDS.

AFRICAN-AMERICANS

We the People Living with AIDS/HIV
(http://www.CritPath.Org/wtp/)

According to the Centers for Disease Control and Prevention, non-Hispanic blacks represent 30 percent of all reported cases of AIDS in the

United States, but make up only 12 percent of the U.S. population.[7] This disproportionate impact has created an urgent need for the most effective, targeted AIDS education and prevention interventions possible, and many authors and producers of AIDS-related Web pages have risen to the challenge. Nearly all of the "best" general HIV/AIDS education and prevention Web sites direct attention to the specific concerns and impacts of epidemic HIV infection in the African-American community (see Appendix). Analytic essays and supporting data can thus be found scattered across many of these Web sites.

Three successful examples, putting the experience of epidemic HIV infection within the black community into context, include the CDC's "HIV/AIDS and U.S. Blacks," available from The Body's Web page (http://www.thebody.com/cdc/factblks.html), "Concern About AIDS in Minority Communities," available at the U.S. Food and Drug Administration's Web site (http://www.fda.gov/fdac/features/095_aids.html), and "What Are African-Americans' HIV Prevention Needs?" provided by the U.S. Center for AIDS Prevention Studies (CAPS) and Harvard AIDS Institute, supported by the University of California, San Francisco (http://chanane.ucsf.edu/capsweb/afamtext.html). What appears to be lacking, however, is a single Web site that collates, evaluates, and provides easy access to the riches available on the Web regarding HIV/AIDS in the African-American community.

One resource that makes a good attempt is the Critical Path AIDS Project-sponsored We The People Web (WTP) site. WTP is a group created by and for people with AIDS (PWAs) and their Web site is used to present a case for an empowered and holistic approach to HIV/AIDS case management. A list of Web sites specifically geared to the African-American community is provided, along with an extensive list of net-based links to important online AIDS resources. According to the WTP Webmasters, "As the largest primarily African American PWA coalition in the nation, We The People also keeps in touch, through the Web, with lots of resources specifically oriented to the African American community, especially the African American sexual minority communities." Links that follow include the Office of Minority Health Resources Center Web page, "A Listing of African American Sites," the "Race, Health Care and the Law" Web page, a listing of "African American Sexual Minority Organizations," and the "National Body of the Black Men's Exchange" Web page, among others. This site, along with the analytic essays, is an important resource for health information professionals, activists, and health care providers to visit when considering the impact of epidemic HIV infection in America's black community.

ASIANS

Asian/Pacific AIDS Resources
(http://www.tufts.edu/%7Estai/QAPA/aids.html)

World infectious disease experts have long been predicting that, next to the African continent, epidemic HIV will take its greatest toll in Asia, where years of denial by many governments and their health officials in nearly every country have led to an appalling lack of access by lay people to HIV/AIDS-related educational and prevention information. The "Asian/Pacific AIDS Resources" Web site is a first step in the right direction. This site includes a section for "Announcements," most recently featuring a call for participants in a national (U.S.) study of Asians and Pacific Islanders living with HIV. The segment, "On-line Resources Asian/Pacific People With AIDS" includes links to the "Asian Community AIDS Services (ACAS) in Toronto," "Critical Path's Asian Language Resources for AIDS," and a link to "Southeast Asian AIDS Resources." A link follows to "Statistics and Analyses," featuring data on AIDS/HIV infection incidence in Thailand, the Philippines, Indonesia, and data current as of 1994 for reported cases of AIDS for Southeast Asia and the Western Pacific. A link is also provided for "Hong Kong AIDS Resources." Following these links is a list of "Support/Resource Groups for Asian/Pacific People with AIDS," representing organizations in Canada, Guam, India, Indonesia, Japan, the Philippines, and the United States, including California, Hawaii, New York, and Washington. Finally, the Web site provides a very brief mediography of films concerning HIV/AIDS in Asian and Pacific Island communities.

This Web site, made available on a Tufts University server, appears to be totally anonymous; no name of a sponsor, organizer, or author appears. Despite this lack of attribution, visitors, particularly individuals impacted personally by HIV infection, will find this Web page an important link in establishing a personal network of resources.

LATINOS AND LATINAS

CLnet: AIDS and the Latino Community
(http://latino.sscnet.ucla.edu/research/aids/aidscomm.html)

CLnet is an Internet-based network service supported by the University of California, Los Angeles, providing a fairly comprehensive online community resource for Latinos. Among the many "Research Center" resources offered by CLnet is the page, "AIDS and the Latino Community." Here you will find links to the proceedings for the 1994 First Binational Conference: AIDS in Our Communities: A Mexico/United States Perspective, proceedings from the 1994 National Latino HIV/AIDS Research Conference, bibliographies on Hispanics/Latinos and AIDS, fact sheets in both Spanish and English about HIV/AIDS prevention, glossaries on drugs and basic terminology also in both Spanish and English, reports on AIDS in Mexico and Puerto Rico, statistical Web-based sites, a listing of Mexican/United States AIDS organizations, and a link to the Hermanos de Luna y Sol AIDS resource page for Latino gay and bisexual men living in the San Francisco Bay Area.

This CLnet Web site is an eminently rich resource for the health professional, researcher, activist, and lay person interested in the impact of epidemic HIV infection in the Hispanic/Latino community. A good number of the links, however, lack obvious authorship, and for the most part much of the statistical data dates from 1994. The overall impression is that this Web site is not being actively updated and maintained. Nonetheless, for the present time, this page provides a unique resource for accessing information where the overriding premise is to link the Latino community with research-based online HIV/AIDS information resources. Let us hope the folks at CLnet realize the value of what they have started.

NATIVE AMERICANS/FIRST PEOPLE

National Native American AIDS Prevention Center: NNAAPC Online (http://www.nnaapc.org/index.html)

The Oakland, California-based NNAAPC is a non-profit corporation governed entirely by Native Americans including people with HIV infection, tribal officials, public health professionals, health care providers, and administrators of substance abuse programs. NNAAPC's mission is to stop the spread of HIV infection among American Indians, Native Alaskans, and Native Hawaiian people. Much of the NNAAPC home page remains under construction, with promised developments under the general categories of "Care" and "Research." The segment on "Prevention," however, does have significant content including an announcement of and an online registration form for a Regional Prevention Training Workshop held in Upstate New York, information on forming Community Prevention Planning Groups, "Technical Assistance" available for regional community-based HIV prevention initiatives, and information on leadership development opportunities for groups working with gay, bisexual and Two-Spirit Native/First People.

Taken as a whole, the resources available at this Web site are clearly directed to activists and impacted individuals looking to develop resources and services to combat epidemic HIV in local Native/First People communities. Don't look here, then, for prevention data or statistics but do visit here if you are looking for how best to organize in your Native/First People community and find support for your efforts at the national level. Visitors to this site can also read an online version of the Center's *In The Wind*, a current affairs bimonthly newsletter on AIDS in Native America, as well as *Two-Spirit Voices*, a newsletter of the National Leadership Development Workgroup for Gay, Bisexual and Two-Spirit Native American Men. Links are provided to a short list of "Native Care: HIV/AIDS Integrated Services Network" resources, including client service and case management models.

RURAL POPULATIONS

Rural Center for AIDS/STD Prevention
(http://www.indiana.edu/~aids/index.html)

According to the U.S. Centers for Disease Control and Prevention, from 1991 to 1995 the number of AIDS cases reported in rural areas of the U.S. increased from 4.9 to 8.8 cases per 100,000.[8] A report in the *Morbidity and Mortality Weekly Report* from 1995 indicates a pattern of HIV infection spread from urban to rural America is well under way.[9] The Rural Prevention Center (RCAP) was established in 1994 to counter this shift.

According to the Center's purpose statement, "The RCAP develops and evaluates educational materials and approaches, examines the behavioral and social barriers to HIV/STD prevention which can be applied to prevention programming, and provides prevention resources to professionals and the public." The Center is a joint project of Indiana University and Purdue University and is funded in part by the Cooperative State Research, Education and Extension Service of the U.S. Department of Agriculture.

Included on the RCAP Web page are links to general Internet AIDS resources, contact information for the RCAP at Indiana University, Bloomington, and RCAP fact sheets dating from 1994 through 1996, covering topics such as "HIV/AIDS in Rural America," "Evaluating HIV/STD Education Programs," "Creating HIV/STD Education Messages for Adolescents," and "Behavior Change Models for Reducing HIV/STD Risk." The RCAP Web site also provides access to the Center's online newsletter, the Spring 1997 issue of which featured articles such as, "Study of Rural Jails Shows Management Procedures May Increase HIV Risk," "Report States that AIDS Prevention Efforts Fail American Young Adults," and "New Curriculum for Migrant Youth." Ongoing RCAP-sponsored projects include "HIV/STD Prevention Education for Rural Students with Special Education Needs," "Development of HIV/STD Peer Educator Training Manual for 4-H Clubs and Other Youth Groups," and "Evaluation of School-based HIV/STD Curriculum in Rural Schools."

The Center supports health care providers, activists, information professionals, and others interested in developing, evaluating, and improving upon their own "home grown" education and prevention services. The RCAP is an exceptional initiative attempting to address a difficult but crucial challenge.

WOMEN

HIVPOSITIVE.COM: Women and Children
(http://www.hivpositive.com/f-Women/WoChildMenu.html)

Education and prevention-related resources about the special and specific impact epidemic HIV infection is having on women, especially women of color, are scattered across the Web, with links to essays, data, and opinion-style articles appearing on general HIV/AIDS information Web pages, and sites dedicated to women's health issues. Such sites of note include the Centers for Disease Control and Prevention's "Factsheet: Women and HIV/AIDS," reproduced on a number of servers on the Internet including the Safer Sex Page (http://safersex.org/women/women.cdc.html); the University of California, San Francisco, Center for AIDS Prevention Studies' "What are Women's HIV Prevention Needs?" Web page (http://www.epibiostat.ucsf.edu/capsweb/womentext.html); and the National Institute of Allergy and Infectious Diseases' "Fact Sheet: Women and HIV" (http://www.niaid.nih.gov/factsheets/womenhiv.htm). One resource that pulls together this disparate data, presents the information as thoughtful narratives, and organizes it all in an easy-to-read fashion is HIV-POSITIVE.COM's Web site dedicated to women and children living with HIV.

HIVPOSITIVE.COM is a commercial Web service marketed to people concerned about HIV/AIDS, including health care practitioners, PWAs, and health information professionals. It is sponsored by TRX Interactive Communications, Inc., New York. The site has been approved by the

Association of Nurses in AIDS Care and honored by the Oncology Nursing Society. The site is actively maintained and updated frequently.

"HIVPOSITIVE.COM: Women and Children" is organized into over ten categories including an overview titled "Women and HIV Infection," followed by "Guidelines for Women with HIV/AIDS," "Female-Controlled Methods for HIV/AIDS Prevention," "Pregnancy and HIV," and "Vertical Transmission of HIV" which considers maternal-child transmission issues. Each category links to a table of contents page featuring further links to a variety of narrative descriptions of the current status of research and information regarding the concept at hand. The overview portion, for instance, addresses general matters of concern to women with HIV infection, including information about transmission, signs and symptoms of infection, symptoms specific to women as regards HIV disease, gynecologic screening, and survival factors for infected women. The succinct and yet thorough narratives are tremendously useful as briefings to current knowledge and understanding about women-specific concerns regarding HIV/AIDS.

According to the editors of this award-winning Web site, "HIV infection is now the third leading cause of death among women ages 25 to 44 and the leading cause of death among black women in this age group." As the demographic profile of epidemic HIV infection continues its apparent shift down the economic ladder, concentrating in minority communities, it will be women of color who will bear the greatest brunt. Women can take heart that the information necessary to save their lives is in fact available and on-hand in online resources such as "HIVPOSITIVE.COM: Women and Children." As health information professionals, we have an obvious responsibility to assist in disseminating and marketing this, as well as all the other, at-risk community-targeted Web sites.

REFERENCES

1. Malinowsky, Robert H., and Gerald J. Perry. *AIDS Information Sourcebook*. Phoenix: Oryx Press, 1988.

2. "AIDS Among Racial/Ethnic Minorities-US, 1993." *MMWR: Morbidity and Mortality Weekly Report* 43 (Sept. 9, 1994): 644-7, 653-5.

3. Biddle, Wayne. *A Field Guide to Germs*. New York: Anchor Books, 1995.

4. Rosenberg, P.S.; Biggar R.J.; and Goedert J.S. "Declining Age at HIV Infection in the United States (letter)." *New England Journal of Medicine* 330 (March 17, 1994): 789-90.

5. Centers for Disease Control and Prevention. *HIV/AIDS Surveillance Report* 5 (1994): 12

6. "Premarital Sexual Experience Among Adolescent Women-United States, 1970-1988." *MMWR: Morbidity and Mortality Weekly Report* 39 (January 4, 1991): 929-932.

7. Karon, J.M. et al. "Prevalence of HIV Infection in the United States, 1984 to 1992." *JAMA* 276 (July 10, 1996):126-31.

8. Rural Center for AIDS/STD Prevention Fact Sheet Number 8 (http://www.indiana.edu/~aids/fact/fact8.html). Bloomington, IN: Rural Center for AIDS/STD Prevention, 1996.

9. Lam, N.S., and Liu, K.B. "Spread of AIDS in Rural America, 1998-1990." *Journal of Acquired Immune Deficiency Syndromes* 7(May 1994): 485-90.

APPENDIX

Following is a select list of some of the best general interest HIV/AIDS education and prevention Web sites currently available.

- AEGIS: AIDS Education Global Information System (http://www.aegis.com)
- American Medical Association. A Physician Guide to HIV Prevention (http://www.ama-assn.org/special/hiv/xi-conf/hivguide/guide.htm)
- Ask Noah About: AIDS and HIV (http://www.noah.cuny.edu/aids/aids.html)
- AVERT: AIDS Education and Research Trust (http://www.avert.org)
- Center for AIDS Prevention Studies (http://www.epibiostat.ucsf.edu/capsweb)
- Centers for Disease Control and Prevention. CDC National AIDS Clearinghouse (http://www.cdcnac.org)
- Centers for Disease Control and Prevention, National Center for HIV, STD and TB Prevention, Division of HIV/AIDS Prevention. HIV/AIDS Prevention (http://www.cdc.gov/nchstp/hiv_aids/dhap.htm)
- Critical Path AIDS Project (http://www.critpath.org)
- HIV InSite: Gateway to AIDS Knowledge (http://hivinsite.ucsf.edu)
- Managing Desire: HIV Prevention Strategies for the 21st Century (http://www.managingdesire.org)
- Peinkofer Associates. AIDS-info.com (http://www.aids-info.com)
- Seattle-King County Department of Public Health. AIDS Prevention Project (http://www.metrokc.gov/health/apu/)
- SIECUS: Sexuality Information Council of the United States (http://www.siecus.org)
- STOP AIDS Project (http://www.stopaids.org)
- The Body: A Multimedia AIDS and HIV Information Resource (http://www.thebody.com)

- TRX Interactive Communications, Inc. HIVPOSITIVE.COM (http://www. hivpositive.com/index.html)
- Vanderbilt University Medical Center, Informatics Center. HIV/AIDS: Education and Prevention (http://www.mc.vanderbilt.edu/resources/interests/aids/edu.html)

Internet Resources
for HIV+ Children and Adolescents

Kris Riddlesperger

SUMMARY. Children and adolescents have unique needs where HIV and AIDS are concerned. They differ from an adult population in terms of education, clinical presentation, and management. This article is a select survey of Internet sites providing information for and about HIV+ children and adolescents. *[Article copies available for a fee from The Haworth Document Delivery Service: 1-800-342-9678. E-mail address: getinfo@haworthpressinc.com]*

The purpose of this article is to provide information to the reader about resources available on the Internet regarding HIV+ children and adolescents.

In the last few years the Internet has become an essential resource for information integral to the understanding of HIV. It is as though the Internet and HIV/AIDS came of age together, providing the consumer who is able to read and use a terminal the latest information on the course and treatment of the virus. Often a patient may obtain new information on the Internet regarding HIV/AIDS, including medications, treatments, fi-

---

Kris Riddlesperger, MS, RN (kriddlesperger@tcu.edu), is Assistant Professor of Nursing at Texas Christian University, Fort Worth, TX. She is completing the requirements for her PhD at Texas Women's University. Her research interests concern HIV/AIDS, and she is co-author, with Jeffrey T. Huber, of a forthcoming book about nutrition and AIDS.

[Haworth co-indexing entry note]: "Internet Resources for HIV+ Children and Adolescents." Riddlesperger, Kris. Co-published simultaneously in *Health Care on the Internet* (The Haworth Press, Inc.) Vol. 2, No. 2/3, 1998, pp. 53-60; and: *HIV/AIDS Internet Information Sources and Resources* (ed: Jeffrey T. Huber) The Haworth Press, Inc., 1998, pp. 53-60; and: *HIV/AIDS Internet Information Sources and Resources* (ed: Jeffrey T. Huber) Harrington Park Press, an imprint of The Haworth Press, Inc., 1998, pp. 53-60. Single or multiple copies of this article are available for a fee from The Haworth Document Delivery Service [1-800-342-9678, 9:00 a.m. - 5:00 p.m. (EST). E-mail address: getinfo@haworthpressinc.com].

nances, and support services. In turn they may share it with their nurses, physicians, social workers, and other professional caregivers, bringing about a collaboration between the patient and caregiver for the treatment of HIV/AIDS. A new and slightly different approach to caring for patients, it is an excellent ally in the progress against HIV/AIDS, working to the favor of both caregiver and patient. A recent example is the advent of the protease inhibitor drugs. As a frequent scanner of the medical sites and support group sites would have observed, protease inhibitors were under discussion several months prior to their actual introduction to the market. Currently, the Internet provides opportunity to trade information about the side effects, reactions, successes, and failures of the medications, thus assisting both patient and caregiver as they work toward an improved state of health. This is just one example of available information about HIV/AIDS on the Internet.

As with many other diseases, children and adolescents have unique needs both when learning about HIV/AIDS and when hosting the disease. Physiologically, the disease progresses on a different timetable in infants and children than in adolescents and adults. Chemically, only five of the eleven approved medications for treating HIV/AIDS are approved for children. Worldwide, of the adults estimated to have HIV/AIDS, 17% have died; but of the children estimated to have HIV/AIDS, 54% have died. Regarding infection rates, the vast majority of infant transmissions of HIV/AIDS are from mother to child either at birth or in utero (and of these the risk of transmission can be decreased from 30% to less than 10% for those mothers taking AZT); but for adolescents the statistic is a startling 25% of overall new HIV cases in the United States. Roughly one adolescent is newly infected every two hours in the U.S. Researchers, health care providers, and educators know that virtually all of these new adolescent cases could have been prevented (Centers For Disease Control and Prevention, 1996. www.cdc.gov).

Awareness of the differences in the disease pathway and numbers like these bring us to the use of the Internet as a source of information gathering and dissemination. It is an invaluable tool in the sharing of resources regarding the needs of children and adolescents learning about prevention, treatment, and living with HIV.

To use HIV/AIDS information on the Internet effectively, three important questions must be asked. First, is the information provided up-to-date? As any Internet "surfer" is aware, often a page is put up only to be left for many months or even years for the general consumption of the public. This is a problem in HIV information. Pages prior to spring 1996 may now present out-of-date information based on the latest advances in medications (protease inhibitors), social services, and attitudes of the general public. There are still a few pages from the late 1980s which present antiquated information regarding HIV/AIDS and are of no utility except as historical overview.

Second, is the information provided accurate? When examining accuracy, one needs to be aware of the source of the page, looking for biases either toward or against children and adolescents who are HIV+. Regardless of the perspective of the page owner, information placed on the page should clearly state whether it is based on legitimate research findings or is the expression of an opinion. This is especially important on pages designed to educate a young population regarding the spread of HIV. Here we enter the murky water of deeply held beliefs regarding sex and young people, what constitutes fact and fiction, and perhaps the role of religion in the treatment and response to HIV/AIDS. A responsible page owner will identify her/himself and the origin of the information presented.

Third, for whom is the page is written? Is it directed for the adult reader or the child/adolescent? Examples of target readers could include teachers who are looking for a curriculum of instruction for adolescent students, parents of an adolescent with HIV, and parents seeking assistance in educating uninfected children about HIV. Or is the page directed toward the child/adolescent her/himself? Differences in direction and target of the page are apparent in the language used and the level of complexity of the information offered.

With these parameters in mind, the following is a summary of visits to pages with information for and about pediatric and adolescent HIV.

PEDIATRIC SITES

Based on the assumption that an adult reading pediatric pages has already taken the time to become acquainted with the basics of HIV and the adult Internet resources, the pediatric pages fall into two primary categories: (1) support, which includes information on emotional, financial, and medical support; and (2) informational, which are informative pages about charities, donations, statistics, and stories about various children (helpful in the desire to put a "face" on HIV).

Talking to Children about AIDS and HIV
(http://www.avert.org/children.htm)

This site is written by and for adults about children. Informative, supportive, and educational, and written in adult language, the page provides a launching pad (many links) to other sites listing children's organizations. It also provides information about talking with children about HIV.

Yahoo!-Health:Diseases and Conditions: AIDS/HIV: Organizations: Children (http://www.yahoo.com/Health/Diseases_and_Conditions/AIDS_HIV/Organizations/Children/)

This is a summary page for adult readers listing 21 children's organizations. The page launches to some unfinished pages, and it is of note that some of the information is quite old. Be aware there are many requests for financial donations on several pages on this listing, but it is a nice summary page.

The Ryan White Foundation: HIV and AIDS (http://www.ryanwhite.org/rwf/hivaids/hivaids.htm)

A great site of support contacts for adults with many offerings of how you can help in the fight against HIV. One of the earlier pages in pediatric HIV (but kept up to date), it calls to mind the courageous fight led by Ryan and his family during a time, in the history of HIV, when society was neither educated nor ready to accept those infected or affected by HIV.

The PediAIDS Electronic News Network Home Page (http://www.hypernet.com/itbic.html)

The PediAIDS Electronic News Network is a super site for pediatric HIV information. Written by adults about children, the mission statement states the page "is designed to provide information and resources for health care providers, educators, researchers, students, the media, and consumers." Sponsored by Boston Children's Hospital and formerly known as In The Best Interests of the Children, this page is a complete resource guide covering or providing links to issues on pediatric HIV and research, programs, camps, medications, and more. An excellent resource, be prepared to spend some time here.

Pediatric HIV/AIDS Links (http://mail.med.upenn.edu/~jstoller/pedaids.html)

Another launch page listing sites specifically of programs for affected and infected children. Written by adults for adults, it is a good resource starting point.

Project Inform Pediatric Resources Hotline Handout
(http://www.projinf.org/hh/ped_refs.html#Books)

Written by adults for the general public, this is a listing of books about pediatric AIDS for children. Listings include books directed toward children with HIV as well as teaching books written at the child level about the virus and AIDS. The site also includes a list of resources with phone numbers for children with HIV.

Healthtouch–Health Information:
Children and Young People with HIV/AIDS
(http://www.healthtouch.com/level1/leaflets/106153/106302.htm)

Provided by a company and written for adults, this is an excellent pamphlet resource center for information on HIV in children and adolescents. The language is simple in many of the pamphlets, and the information is up-to-date and reliable.

National Children's Coalition
(http://www.child.net/ncc.htm)

This site, called Kidcity, is included because it is designated to be by kids for kids (with a little adult help). If the promises made on the cover page are fulfilled, it will be an exciting site for children with HIV, and for adults as well.

This is not, by any means, an exhaustive listing. There are a significant number of other pediatric HIV sites which are worth looking into, provided one has the time and inclination to do so.

ADOLESCENT SITES

Adolescent sites abound for the education of the young population regarding HIV. There are sites written for teens by teens, for teens by adults, and for teens by health providers covering all ranges of the spectrum of adolescent information. However, most of the medical information is found under the adult pages in general information about HIV/AIDS. Parents should be aware that occasionally, when a page is written for teens, the language often is in "teenspeak," employing graphics and utilizing common terms for body parts and activities. While some may find this offensive, most will agree it is acceptable as long as it is accurate, promotes the education of an adolescent, and gets the message across.

Eastchester Middle School: The AIDS Handbook
(http://www.westnet.com/~rickd/AIDS/AIDS1.html)

This is a site prepared by Eastchester middle school students for middle school students (11-14 years old) about HIV/AIDS. It is an excellent site and has been selected as an Internet resource for the Discovery Channel School's Body Science theme for fall 1997. This page reflects a great deal of hard work on the part of the students and faculty and is an excellent, age-appropriate place for a young teen to begin education about HIV. The page also includes links to other adolescent and general information sites regarding HIV and education.

HIV/AIDS: What's This Got To Do With Me?
(http://members.aol.com/ananiu/aids.htm)

Using a menu format, this page is written for teens in appropriate but often blunt language. Available for your perusal are stories written about and by teens with HIV; a letter from a physician who is an infectious disease specialist and has been treating young people with HIV since 1983; and serious, specific discussion about safe sex, safer sex, and safest sex. This page has a sense of humor as the language becomes most specific, but the information is accurate and frank. The reader would not leave this page wondering about definitions or terms. An interesting additional feature of this page is a Q and A spot by Dr. Ruth. Her topic on 9/13/97 was directed toward providing information about oral sex and the incidence of transmission of HIV via this route. The page includes links to other good sites for further information on numerous topics in HIV/AIDS.

CAPS Hot Topic: Adolescents
(http://www.epibiostat.ucsf.edu/capsweb/hotteens.html)

CAPS (Center for AIDS Prevention Studies from UCSF) sponsors this site for adolescents with discussions of sex, HIV, and teens. There is information here for teens about talking to your parents concerning HIV and sex, and for parents regarding how to talk to your teen about HIV and sex. This is an excellent site for education and an update on adolescents and HIV.

AIDS Now! For Teens
(http://itec.sfsu.edu/aids/aids.html)

This site is a five-question quiz for teens. If all questions are answered correctly, the student may print out a certificate of congratulations. The site specifically asks for the name of the person taking the quiz. The questions are:

- What is AIDS?
- What is the cause of AIDS?
- How can you get infected with HIV?
- What activities are HIV risk free?
- Why is AIDS so dangerous?

The answers are in multiple choice format. The reader is referred to further information if a question is missed for the correct facts. When the author intentionally missed a question, the information received did not seem appropriate for a young teen, although it was fine for an older adolescent. As with most other pages, resource links are available from this page for further HIV information.

HIV and AIDS Information for Young People, Including Safe Sex (http://www.avert.org/young.htm)

This page, titled AIDS Information For Young People, provided by the AIDS Education and Research Trust of England, is self-described as "delivering some information that we have found young people want to know." The information provided addresses what is AIDS, what causes it, how it is contracted, what is safe sex, and the notion of "it won't happen to me." This is a very basic, beginner level page in easy-to-understand language. Very simple and nice for the younger adolescent, but an understanding of terms is needed before accessing the reading material.

The Park School AIDS Awareness Committee Home Page (http://pages.prodigy.com/parkaidsaware/)

Provided by The Park School AIDS Awareness Committee, this site is specifically devoted to adolescents age 14 to 18. The facts about HIV are presented. True stories about teens living with HIV are provided, but the most interesting idea on this page is the invitation for youth to post an article on the site for general discussion. Although articles are not always available, this is a great method to invite participation and discussion from the target audience. Links to other informational sites for teens are accessible from the site. The only questionable note about the site is they still have a brochure on the page about a conference held January 1997, which leads one to wonder about their ability to remain updated in the rapidly changing arena of HIV. Thus far, however, the information provided is up to date.

Youthco AIDS Society Homepage
(http://www.youthco.org/)

Sponsored by the LesBiGay.com group of Web sites, the Youthco mission statement reads: "Youthco is a peer-driven organization based in Vancouver, British Columbia. It strives to enable youth from all communities to address youth issues concerning HIV/AIDS by acting as a resource and facilitator for educational initiatives and support." This is a support page providing information from an educational as well as alternative lifestyle point of view. It is blunt in language, but appropriate to the audience it is addressing. The page, with some additional refinement, can meet a specific need in the population of adolescents addressing issues in HIV. However, at present, it is a bit rough around the edges.

CONCLUSION

Much more information is available on the Internet regarding children and adolescents than listed in this brief review. The reader is encouraged to decide what type of information is needed and to then pursue that area in depth following the offered links from pages of similar interests. Maintaining an awareness that a site may be out of date, present an idea or value with which you disagree, or employ language different than that with which you are familiar, will serve to enhance your search and refine it to pages of special interest to you and your child or adolescent.

Women's Place on the World-Wide Web:
An Analysis of Sites
Concerning HIV and Women

Mary L. Gillaspy
Jeffrey T. Huber

SUMMARY. The Internet contains hundreds of sites specific to HIV and AIDS; however, few of them are specific to women, the fastest-growing group of infected individuals worldwide. Since women's experience of HIV disease differs significantly from men's and their prevention needs are different as well, a need for female-specific Internet information exists. Most such information currently available is part of large, well-established sites. This article reports a partial survey of such sites, as well as others targeting special populations of women. Development of and access to eventual sites is complicated by the global demographics of female infection. However, electronic sources of information concerning women and HIV are necessary not only for infected individuals but also for prevention and for people who work with either population. *[Article copies available for a fee from The Haworth Document Delivery Service: 1-800-342-9678. E-mail address: getinfo@haworthpressinc.com]*

Mary L. Gillaspy, MS, MLS (gillaspy@utmdacc.mda.uth.tmc.edu), is Manager, The Learning Center, The University of Texas M.D. Anderson Cancer Center, Houston, TX 77030. Jeffrey T. Huber, PhD (jeffrey.huber@mcmail.vanderbilt.edu), is Research Information Scientist, Active Digital Library, Eskind Biomedical Library, and Research Assistant Professor, Division of Biomedical Informatics, School of Medicine, Vanderbilt University Medical Center, Nashville, TN.

[Haworth co-indexing entry note]: "Women's Place on the World-Wide Web: An Analysis of Sites Concerning HIV and Women." Gillaspy, Mary L., and Jeffrey T. Huber. Co-published simultaneously in *Health Care on the Internet* (The Haworth Press, Inc.) Vol. 2, No. 2/3, 1998, pp. 61-78; and: *HIV/AIDS Internet Information Sources and Resources* (ed: Jeffrey T. Huber) The Haworth Press, Inc., 1998, pp. 61-78; and: *HIV/AIDS Internet Information Sources and Resources* (ed: Jeffrey T. Huber) Harrington Park Press, an imprint of The Haworth Press, Inc., 1998, pp. 61-78. Single or multiple copies of this article are available for a fee from The Haworth Document Delivery Service [1-800-342-9678, 9:00 a.m. - 5:00 p.m. (EST). E-mail address: getinfo@haworthpressinc.com].

INTRODUCTION

Though rarely acknowledged equally with their male counterparts, women have been statistics in the HIV/AIDS pandemic since its inception, both in the United States and in the rest of the world. The epidemiology of HIV disease and its presentation, barriers to care, and prevention efforts differ significantly based on gender.

Information about HIV specific to women has historically been sparse, despite the huge body of scientific literature concerning the malady. However, as the number of infections in U.S. women has risen, so have the number of journal articles and books that either include or are devoted to female persons living with AIDS (PLWAs). The growing amount of data on the Internet, and access to it, represents another possible avenue for the dissemination of information about women and HIV/AIDS. Though comparatively few people in the United States, much less the world, currently have access to the Internet, it remains a viable channel of information, one likely to become ever more pervasive in a very short period of time. Moreover, the diverse grass roots responses to the HIV epidemic in the United States have made Internet access widely available through community-based venues, at least in urban areas. Access by health care workers on behalf of patients, as well as connections through public libraries and community-based organizations, also permit HIV-infected women to acquire knowledge about their disease process, therapeutic choices, and the overwhelming psychosocial challenges they face.

BACKGROUND

Epidemiology

In 1981, when the Centers for Disease Control and Prevention (CDC) first began tracking AIDS cases in the United States, women represented 3% of the total reported. By the end of 1996, that figure had jumped to 20%.[1] The number of women reported infected with HIV but who had not progressed to AIDS in 1996 in the United States was even higher–30%.[2] Worldwide, 8.8 million women are estimated either to be infected with HIV or to have AIDS. By the turn of the century, epidemiologists estimate that the numbers of men and women worldwide infected with HIV will be equal.

The majority of all AIDS cases in the United States are found in the Northeast, so it is not surprising that most cases among women are found there as well (44% of the total). The South contains the second largest percentage of female cases, 36%. However, while nearly all of the female cases in the Northeast are found in or very near urban areas, more than 10% of the infected women in the South and the Midwest live in rural areas.[3] Ethnic statistics show that new infections are occurring overwhelmingly among women of color, African-Americans and Hispanics. Though African-American and Hispanic women comprise a mere 21% of all females in the United States, the rate of reported cases of AIDS in these two groups has ranged from 75% to 78% of all new cases among women over the past five years.[4,5,6,7] These demographics, coupled with the common variable of poverty in the cohort, indicate the magnitude of the challenge facing anyone attempting to provide life-saving information to these women. The Internet represents one of several important communication channels that can be used for this purpose.

Tracking the cases among women also demonstrates the relentless march of HIV into the heterosexual population in the United States. Though the cases among heterosexuals are not increasing exponentially the way they did among the gay male population in the 1980s, the trend toward increasing transmission via heterosexual contact is clear. In fact, of all the AIDS cases currently occurring in women between the ages of 13 and 24, nearly half (49%) are attributable to heterosexual transmission. Epidemiologists at the World Health Organization (WHO) predict that heterosexual transmission will soon be the dominant means of transmission in North America and Europe, just as it has always been in Africa and Asia. Certainly this appears to be the trend in the United States. In 1995, for the first time, more AIDS cases reported among women were transmitted heterosexually than via injection drug use. The trend was even more pronounced in 1996 (see Table 1). Yet another trend is in increased mortality rates among women. At the same time as deaths among men with AIDS have declined 15%, deaths among women have increased 3%. To put the numbers into perspective, consider that worldwide, two women

TABLE 1. Transmission of HIV Among U.S. Females, 1992-1996*

	1992	1993	1994	1995	1996
Injection Drug Use	47%	47%	41%	38%	34%
Heterosexual Contact	39%	37%	38%	38%	40%
Total	86%	84%	79%	76%	74%

*Between 13% and 24% of cases appear as "Risk not reported or identified."

every minute of every day become infected with HIV, and every two minutes of every day, a woman dies of AIDS.[8]

At a time when much of the public seems to believe that HIV has become a treatable condition, since the overall mortality rate has declined, the widest possible provision of information is key to slowing the spread of the disease. Ann Duerr, Chief of the CDC's HIV Section, Women's Health and Fertility Branch, concludes, "The AIDS epidemic [in the United States] has not yet reached a plateau for women, but it has plateaued for men."[9] Whether or not one agrees that the epidemic has truly peaked in any population, some segments of society continue to think of HIV as a disease affecting persons other than themselves,[10] presumably gay white men and injection drug users. Internet sources can play an important role in educating that portion of the public that believes itself immune from AIDS.

Patient Presentation

Typically, HIV-infected women present at a different stage in their disease process and with different symptoms and, to some extent, with different opportunistic infections from men. Current data suggest that the HIV virions themselves differ from the outset in men and women. Initially, a single virion generally appears in newly infected men; in the same cohort of women, however, heterogeneous virions are evident.[9] While further study will be necessary to explain this finding, the greater variety of virions in women from the onset of infection suggests a possible reason why women appear to get sicker faster and to show resistance to anti-retrovirals earlier than their male counterparts.

Of course, most of the conditions, infections, and neoplasms unique to women are gynecologic in nature. Invasive cervical cancer is as telling a sign of HIV infection in young women as Kaposi's sarcoma is in young men. Unusual types of malignant breast tumors are also appearing among women with HIV. Cervical intraepithelial neoplasia (CIN), pelvic inflammatory diseases (PID), and vulvovaginal candidiasis represent serious problems in immunocompromised women. Women are more likely than men to present with esophageal candidiasis, a condition associated with malnutrition and HIV wasting syndrome. In women as well as men, prior infection with any human papilloma virus (HPV) not only increases the likelihood of sexual transmission of HIV but also presents serious medical management problems within immunocompromised individuals.

Most importantly, access to early care, with aggressive treatment in the early stages provided by a physician who specializes in HIV disease, has been shown to be key to increased survival of all infected persons.[11] "The majority of women with AIDS and HIV are relatively young, come from an impoverished background, and represent minorities,"[9] all factors which mitigate against access to medical care, especially by specialists. But when gynecologic manifestations do not respond to usual treatments, both women and their health care providers should suspect HIV disease and test for it. The Stadtlander World-Wide Web site (http://www.stadtlander. com/hiv/hivwomen.html) discussed later in this article outlines the differences in presentation between gynecologic maladies in HIV-positive and HIV-negative women.

Barriers to Care

The first barrier that HIV-infected women have always encountered is that they are not men. Especially in the early years of the epidemic, men, especially gay and/or injection drug-using men, formed the majority of the AIDS caseload. Many women who actually were infected were not even counted in the official statistics,[12] partially because they did not suffer from one of the AIDS-defining conditions recognized at that time. Indeed, not until 1993 did the CDC list invasive cervical cancer as an AIDS-defining condition; by this time, the agency had counted 25,202 AIDS-related female deaths. Throughout the years of the epidemic, women have routinely presented when they were sicker than men, so their survival rate from date of diagnosis has been lower than their male counterparts.

In society generally, women are less economically well-off than men; certainly this is true in the HIV+ community. HIV disease is extremely expensive over the course of the patient's life; most infected women simply have not been able to afford the care that would keep them alive, even if they have had access to it. Currently, when protease inhibitors (PIs) are extending the lives of many patients, states are restricting access to these life-saving drugs or capping the total amount they will pay per patient for all HIV care, if individuals do not have private insurance or cannot personally pay the costs. Indeed, only New York and North Carolina cover the costs of all the antiretrovirals and protease inhibitors presently available; four states cover none of the PIs.[13]

Access to clinical trials represents a perennial barrier, not only for HIV disease but for other health care problems as well. The protease inhibitors, currently the most promising of all anti-HIV drugs when used in combination with other substances, have been tested primarily on white males. Just as men did before them, some women are attempting to gather their own data and disseminate the results electronically. The *Women Alive* newsletter (http://www.thebody.com/wa/jan97/survey.html) has distributed a six-page questionnaire and is asking women taking PIs to complete it and mail it in. The introduction to the questionnaire states, "This survey came about because of the need for answers based on facts, not speculation, and because of the serious lack of useful information for women about the new class of drugs." Results of the survey have not been published as of this writing. A promising development noted at the 3rd National Conference on Women and HIV, held in May 1997, is that in studies funded by the AIDS Clinical Trials Group (ACTG), the proportion of HIV-infected women to men enrolled has increased from 5.6% in 1987 to 16.9% in 1996,[9] just over half of the documented number of cases. Since women currently comprise one-fifth of the AIDS caseload in the United States, the percentage of enrollees may increase.

Finally, ever since the medical and research communities acknowledged women's presence in the ranks of HIV-positive individuals, the focus has seemed to be more on women as vectors of transmission to a fetus than on the women themselves. Even less fair was the emphasis on female commercial sex workers' being vectors of transmission to their male clients, rather than the other way around.[14] Mandatory testing has been discussed more in terms of pregnant women and commercial sex workers than in any other group besides prisoners and members of the military. Yet, male-to-female transmission of HIV is approximately 2.3 times more likely than female-to-male transmission.[15]

Owing to the documented decrease in perinatal transmission of HIV if pregnant women take Zidovudine (AZT),[16] some politicians and pundits urge the mandatory testing of all pregnant women. However, mandatory testing of anyone is more problematic than it may first appear. Women may be in relationships that could result in abuse or abandonment if their HIV seropositivity became known to their partners. Since the majority of women infected are young, poor, and Hispanic or African American, they may not relate to the health care system in the same way that middle class Anglos do. For reasons not yet known, women are less likely than men to be compliant with medication regimens, resulting in a greater chance of drug resistance and the possibility of transmission of resistant strains of HIV.[9] Some activists fear that if one group is subjected to mandatory testing, the practice will spread, raising the spectre of a requirement for some means of visual identification or quarantine of HIV-infected individuals, practices widely discussed in the 1980s.

The preference and accepted practice as of this writing is for health care providers to educate all women about their risk for infection, to counsel them as to their options, and to encourage voluntary testing, following strict guidelines for informed consent, for all pregnant women and those who are considering pregnancy. One of the pages on the Health Care Financing Administration (HCFA) site is entitled Doctors: It's Your Responsibility (http://hiv.hcfa.gov/subpg5.htm) and states categorically: "All health care providers who treat pregnant women (or women thinking about becoming pregnant) need to test their patients for HIV. Currently, physicians are being urged to offer voluntary HIV testing of pregnant women; however, in the future, this testing may become mandatory." Yet, since the women most likely to be seropositive are also most likely to be poor women of color, they are unlikely to have full access to appropriate medical interventions for either themselves or their children. The issue remains controversial, not least for the fact that women and their welfare appear secondary to a fetus that may not yet even be conceived.

No amount of information can overcome all the barriers to care that currently exist. However, for women disposed through whatever means to educate themselves about their risks, Internet access can help them determine their options, provide them with psychosocial support, and, if they test positive, empower them to live with HIV, not be a victim of it.

Prevention Needs

As is true of epidemiology, presentation, and barriers to care, the prevention needs of women differ significantly from those of men. In particular, it is important to reach adolescent women who may already be sexually active or thinking about becoming so. Among this group, Internet interventions hold great promise for efficacy. Most of the listed sites contain actual scripts that women can use to negotiate safer sex with a partner. Such techniques are increasingly viewed as preferable alternatives to Reality, the female condom. When it was introduced, Reality was seen as a great advance for both contraception and STD prevention, since women control its use. Among the female population most at risk for HIV infection, however, its use presents challenges. It is much costlier than male condoms, and it requires education to be used effectively. The scripts published on the Internet may assist women to assert the need for safer sexual practices with their partners and thus protect them from infection, as Reality is intended to do. Redundant approaches to prevention, particularly in a population that perceives itself to be at low risk, are a correlate to the combinations of factors heterosexuals indicate are reasons they do not practice safer sex.[10]

The 3rd National Conference on Women and HIV focused on prevention, since it remains the only sure way for sexually active men and women to avoid infection.[17] Four recommendations for research and product development in the area of prevention emerged from the proceedings:

1. Maintain and restore, if necessary, the natural protection against all sexually transmitted diseases (STDs), including HIV, present in the normal vagina;
2. Develop improved prevention tools that will be more acceptable to both men and women than the technologies currently available;
3. Emphasize prevention, screening, early detection, and treatment for sexually transmitted diseases other than HIV;
4. Perform prevention research to learn more about female anatomical risk factors for HIV.[3]

This clearly outlined research agenda may serve as a guideline for creating and evaluating Web sites, and is a logical complement to excellent sites that focus on prevention. The Internet represents only one tool that health educators, health care providers, and community workers employ to teach and encourage safer sexual practices among women. However, particularly among young women and students, it can be a powerful educational tool to facilitate behavior change that may result in a healthier population.

EVALUATIVE CRITERIA

Sites included were evaluated according to four broad criteria: scope, content, workability, and design. *Scope* includes both breadth and depth; *content*, accuracy, authority, currency, uniqueness, links, quality of original material, and reading level; *workability*, user friendliness, search engine (if available), and logical organization; and *design*, text and graphics. The sites included are not meant to be a comprehensive listing of all that are available on the Internet. Rather, the list represents major sites that offer a significant amount or quality of information to or about women and HIV/AIDS as well as a few sites devoted just to women and their HIV concerns. They are divided into six categories: general sites with information for women; pregnancy issues; special populations; women and prevention; personal stories and art; and indexes, pages of organized links, and encyclopedic information. Not included are U.S. regional sites from the many community-based organizations (CBOs) across the country. However, many of the larger sites include links to CBOs, and they can certainly be easily accessed by women living in specific areas.

DESCRIPTION OF SITES

General Sites with Information for Women

Stadtlanders: HIV Infection in Women
(http://www.stadtlander.com/hiv/hivwomen.html)

This article, authored by Marion Banzhaf, coordinator of the New Jersey Women & AIDS Network, focuses on the gynecologic signs of HIV seropositivity, signs which are often ignored by both women and their health care providers. The focus on undiagnosed HIV infection may lead readers of this page to testing, thus prolonging their lives if infected.

Pos+Net on the Internet
(http://www.posnet.co.uk/public/default.htm)

This site from the United Kingdom includes one section devoted to women. The links are all to pages in the United States, but a bulletin board service (BBS) provides the opportunity to interact with other HIV-positive individuals. A discussion list is also available.

Project Inform Fact Sheet Packet: Guidelines for Women with HIV/AIDS 1
(http://projinf.org/fs/women1.html)

In keeping with Project Inform's mission, this article provides carefully researched, well-written information about social, psychological, and sexual concerns; physical concerns; and pregnancy and resources. Comprehensive treatment information for specific gynecologic abnormalities is provided.

Project Inform Fact Sheet: Guidelines for Women with HIV/AIDS 2
(http://www.projinf.org/fs/women2.html)

This reprint from *AIDS Clinical Care* (1996) addresses the nervous, endocrine, and immune system interactions in HIV-positive women and what is known about them. Questions about menstruation, menopause, hormone replacement therapy, and the use of oral contraceptives among HIV-infected women are addressed, and current knowledge on these topics is summarized.

JAMA HIV/AIDS Information Center
(http://www.ama-assn.org/special/hiv/hivhome.html)

This outstanding general site provides quick and easy access to the AIDSDRUGS, AIDSLINE, AIDSTRIALS, and BIOETHICSLINE databases.

HIV InSite: Key Topics: Women
(http://hivinsite.ucsf.edu/)

From the home page, click on *Key Topics*. One entire area is devoted solely to women, including women who have sex with women. Parts of some of the other key topics, like adolescents, the ethnicity-specific topics, and children are also applicable to women and HIV. The frequently asked questions (FAQ) documents are particularly valuable and easily available from this site.

✂ Pregnancy Issues

Pregnancy and HIV: What Women and Doctors Need to Know
(http://hiv.hcfa.gov/)

This well-done, simply written Web document includes information about the Maternal AIDS Project, Medicaid coverage for pregnant women, testing for HIV, and how to reduce the possibility of perinatal transmission. A half-page handout is entitled "Pregnant or Thinking About It" and is remarkable for its nonjudgmental tone and ease of use in a variety of settings.

See also: HIV InSite, JAMA HIV Information Center, Critical Path AIDS Project, AIDS in Mexico and the World, and Women Alive.

Special Populations: Speakers of Languages Other Than English

AIDS in Mexico and the World: An Electronic Library Without Borders
(http://jeff.dca.udg.mx/sida/Ingles/aids.html)

This site is bilingual, English and Spanish. The emphasis is on pregnant women, but there is a great deal else here that is easily accessible in either language. (The Spanish version is better written than the English translation.) A simply written family information guide explains basic HIV facts and ends with the statement, "You can live with an infected person or a person already sick with AIDS without danger to yourself. Give that person care and understanding. Do not abandon them." A quick time video is included, as is a page about how to increase condom effectiveness. The page devoted to women is entitled "What Every Pregnant Woman Should Know About HIV and AIDS: What You Can Do to Have a Healthy Baby" and comes from the Pediatric AIDS Foundation. This site is most useful to speakers of Spanish.

VIH y SIDA
(http://www.ctv.es/USERS/fpardo/home.html)

This is another bilingual English/Spanish site, but from Spain, so the perspective is European. See also the New York Online Access to Healthcare, or NOAH.

ABIA
(http://ibase.org.br/~abia/)

ABIA, an acronym for Associacao Brasileira Interdisciplinar de AIDS, or An Interdisciplinary Brazilian Association of AIDS, is comprehensive and is constructed almost entirely in Portuguese. Interestingly, the English acronyms HIV and AIDS, rather than their Portuguese equivalents, are used throughout the pages.

Migrants Against AIDS/HIV (MAHA)
(http://www.hivnet.ch/migrants/stand.html)

Originating from Geneva, Switzerland, this site is partially bilingual in English and French. MAHA contains unique information because of its European base and its focus on refugee, immigrant, Black, and Third World communities in Europe. The site consists primarily of a description of the organization and its newsletter, many issues of which contain information for and about women in the targeted communities. However, the issues of these women are the issues of women all over the world. For example, one article editorializes concerning the French Academy of Medicine's recommendation that HIV testing be made mandatory for all pregnant women. While this site does not target women per se, it is worth watching for women's issues within this singular socio-politico-cultural context.

Special Populations: Rural Women

Rural Center for AIDS/STD Prevention
(http://www.indiana.edu/~aids/)

Developed and maintained jointly by Indiana and Purdue Universities, this is an important resource for a population often forgotten in the urban AIDS statistics. One of the fact sheets, "HIV Infection and Women," addresses incidence, risks, differences in the disease between men and women, testing, and prevention. A fact sheet on "HIV/AIDS in Rural

America" focuses on some of the special problems of HIV-infected rural populations and on the challenges of HIV prevention outside of large cities.

Special Populations: Adolescents

Coalition for Positive Sexuality
(http://www.positive.org/cps/Home/index.html)

The title of this site may be misleading. It uses the word *positive* in the pre-AIDS sense of the term. The pages are written for "teens who are sexually active now or just thinking about having sex." The FAQ is bilin-gual, English and Spanish. The content is straightforward, and sexually transmitted diseases (STDs) other than HIV are also addressed unambiguously. See also HIV InSite's Key Topics: Adolescents.

Special Populations: Women Only

Women Alive
(http://www.thebody.com/wa/waix.html)

Part of the encyclopedic *The Body* site, *Women Alive* contains the full text (searchable) of this organization's quarterly newsletters, dating from Autumn 1993. Some articles from the newsletter are in Spanish, but it is not really a bilingual publication. Women Alive, based in Culver City, California, operates a national hotline and provides peer support and counseling for women living with HIV and AIDS. The newsletter focuses on treatment (traditional as well as alternative) but also discusses psychosocial concerns and includes an area for infected women's children to describe their experiences. The January 1997 issue includes a survey for women taking protease inhibitors, so that their experiences with these powerful drugs can be monitored and recorded.

Yoohoo! Lesbians!
(http://www.sappho.com/yoohoo/)

This site covers many areas specific to lesbian life; the URL provided here takes users directly into the Health & Sexuality and Lesbian-specific Health Issues areas. Homophobia among health care providers is discussed, as well as particulars of both physical and mental lesbian health issues. "We Are Not Immune" addresses all STDs and explains how to avoid contracting or transmitting them.

Women and Prevention

Most of the best prevention information on the Internet is a subset of another site. One site, however, stands out from all the rest because of its emphasis on negotiation for safer sexual practices between partners, respect for oneself and one's partner, and factual information, simply written and illustrated. That site is the following one:

Condomania
(http://www.condomania.com/)

The segment of *Condomania* entitled "Getting Educated" includes scripts for negotiation (non-gender-specific); an inventory of which practices are considered low risk, medium risk, and high risk; tips on testing; a listing of resources and hotlines; and much more.

Following are descriptions of prevention segments of larger sites that may be particularly helpful to women and to health educators working with women in high-risk categories.

Project Inform Fact Sheet Packet: Guidelines for Women with HIV/AIDS 1
(http://projinf.org/fs/women1.html)

This site includes advantages and disadvantages of the Reality condom and a straightforward listing of activities that pose no risk, some risk, or high risk of HIV transmission.

HIV InSite: Key Topics: Women
(http://hivinsite.ucsf.edu/)

Nine HIV Prevention Fact Sheets, all but one of them bilingual, address either women directly ("What are Women's HIV Prevention Needs?" and "What Are Women Who Have Sex With Women's HIV Prevention Needs?") or groups of which women represent a significant share of the total (adolescents, prisoners, African-Americans, Latinas, homeless people, commercial sex workers, and substance abusers). The Fact Sheets were developed and are maintained by the Center for AIDS Prevention Studies at the University of California, San Francisco.

Critical Path AIDS Project
(http://www.critpath.org/)

The portion of Critical Path AIDS devoted to women is particularly strong in the area of lesbians' and bisexual women's prevention needs. It includes a "Lesbian Safer Sex Guidelines" cut-and-fold display and links to a variety of other sites devoted to women and women's health interests.

The Body
(http://www.safersex.org/)

In keeping with the global nature of this resource, this segment of *The Body* contains links to many other prevention sites as well as some original information.

Gay Men's Health Crisis
(http://www.gmhc.org/stopping/disclaimer/disclaimer.html)

This site contains the most simply written prevention information of all those surveyed. Under the heading "Women Need to Know About AIDS," the pages discuss who, why, and how, as well as female-specific information about pregnancy, children, artificial insemination, and negotiation. GMHC is also the home of the Lesbian AIDS Project, which originated much of the prevention information specific to this population.

Personal Stories and Art

The ephemeral nature of Internet information is nowhere more evident than in the category of *Personal Stories and Art*. Two large sites that contain an example of each at this writing are POZ (http://poz.com) and the Coming Into View: Women, Families & AIDS (http://www.aidsnyc. org/berridge/index.html). The POZ site contains profiles of various individuals, including HIV+ women; the Coming Into View site includes photographs and personal stories of HIV-infected women, all the work of Mary Berridge. The latter site is particularly useful in showing the diversity of HIV+ women.

Indexes, Organized Links, and Encyclopedic Information

HIV Info
(http://www.info.org/)

This index of HIV/AIDS information available on the World-Wide Web includes direct links to chat lines and a Monthly Political Action Statement that facilitates individuals' contacting power brokers in both the private and the public sectors concerning various HIV issues.

The HIV/AIDS Project of the Active Digital Library,
Vanderbilt University Medical Center
(http://www.mc.vanderbilt.edu/adl/aids_project/internet.html)

This carefully indexed site contains both local and global resources. Electronic resources include those from the National Library of Medicine, the National AIDS Clearinghouse, various medical databases available via the Internet, and Internet resources. This last category is divided into basic science, conferences and meetings, education and prevention, epidemiology, information resources, medical management, news and discussion groups, organizations and projects, and psychosocial and religious issues. Users may e-mail questions regarding information sources and resources into the site and receive a response from an HIV/AIDS specialist.

AEGIS (AIDS Education Global Information Center)
(http://www.aegis.com/)

The Centers for Disease Control and Prevention call AEGIS the best of its kind. Indeed, this encyclopedic resource, operated from 1990-1995 by the Sisters of St. Elizabeth of Hungary, advertises itself as the largest HIV/AIDS database in the world. AEGIS is now a nonprofit organization, but its guiding light and master is still Sister Mary Elizabeth, its founder. AEGIS offers reliable access to the CDC's *AIDS Daily Summary* and to a search engine that seamlessly accesses the huge variety of information in the database, much of which is specific to women.

Critical Path AIDS Project
(http://www.critpath.org/)

Another encyclopedic resource, this site is notable because it was founded and continues to be operated by persons living with AIDS (PLWAs) and so is organized with patients' needs in mind.

New York Online Access to Health (NOAH)
(http://www.noah.cuny.edu/aids/aids.html)

The great advantage of this site for all health topics is that it is fully bilingual, English and Spanish. It offers easy access to the brochures and fact sheets on opportunistic infections from both the Gay Men's Health Crisis and the National Institutes of Health. An entire area of the index is devoted to "Women, Youth, and Race Issues." If basic information in Spanish is desired (other than treatment information), this site is the place to begin.

The Body: A Multimedia AIDS and HIV Information Resource
(http://www.thebody.com/cgi-bin/body.cgi)

This excellent site divides HIV/AIDS information into the basics, treatment, conferences, quality of life, and government. An exceptional feature is the "Insight From Experts" section.

AIDS Treatment Data Network
(http://www.aidsnyc.org/network/index.html)

ATDN and Project Inform are two of the best places to begin looking for treatment information. ATDN's "Simple Fact Sheets" are simply written, one-page documents, available in both English and Spanish, that cover not only treatments but also specific conditions and various tests, such as the polymerase chain reaction (PCR).

DISCUSSION

This listing of Internet sites concerning women and AIDS is meant to be neither comprehensive nor representative of every site available. Of the many Internet sites operating at this time, the ones included in this article seem particularly useful for women seeking HIV information, whether or not they are infected. Since the numbers of infected women are unfortunately growing, the need for continued development of quality sites that focus on management and treatment issues specific to women is obvious. However, such development can only occur following or concurrent with biomedical research that examines women's therapeutic and medical management needs in the context of the epidemic. Finally, the largest demographic groups of infected or at-risk women are barely addressed in existing sites, if they are discernible at all. Sites targeting injection drug-using women, sexual partners of injection drug users, married women, and others are badly needed, together with community-level access to the electronic information for these groups.

CONCLUSION

It can be argued that since many excellent HIV/AIDS sites already exist on the Internet, it should not be necessary to differentiate the information for women. However, the infected female population differs significantly from their male counterpart. Even though the major risk factor for infection among women is heterosexual contact, most heterosexuals, men and women, do not consider themselves at risk for HIV.[18] Women's experience of HIV disease differs from men's, and the bulk of infections occur among poor women of color. Most existing sites target upper middle class, well-educated individuals and may not be culturally accessible to the majority of infected women.

In light of the demographics of HIV-infected women, perhaps HIV information on the Internet that targets women will make the most demonstrable difference among young women ignorant of their risk (especially adolescents), caregivers and health care workers providing services and information to infected women, and the general public. But, particularly as opportunities for access to electronic information sources expand, some HIV-infected women will find their way to Internet sources designed especially for them and will acquire the information and knowledge they need to care for themselves and make decisions about their own lives and the lives of their children.

REFERENCES

1. Centers for Disease Control and Prevention. HIV/AIDS Surveillance Report 8 no. 2 (1996): 10.

2. Centers for Disease Control and Prevention. HIV/AIDS Surveillance Report 8 no. 2 (1996): 30-31.

3. Hanna, Leslie. "Selected Highlights from The National Conference on Women & HIV, Pasadena, CA, May 4-8, 1997 (http://library.jri.org/library/news/beta/beta33h.html)." BETA: Bulletin of Experimental Treatments for AIDS 33 (June 1997).

4. Centers for Disease Control and Prevention. HIV/AIDS Surveillance Reports 8 no. 2 (1996): 12.

5. Centers for Disease Control and Prevention. HIV/AIDS Surveillance Reports 7 no. 2 (1995): 13.

6. Centers for Disease Control and Prevention. HIV/AIDS Surveillance Reports 6 no. 2 (1994): 13.

7. Centers for Disease Control and Prevention. HIV/AIDS Surveillance Reports 5 no. 2 (1993): 10.

8. Stine, Gerald J. AIDS Update 1997. Upper Saddle River, NJ: Prentice Hall, 1997.

9. "No Plateau for HIV/AIDS Epidemic in US Women." JAMA: The Journal of the American Medical Association 277 (June 11, 1997): 1747-9.

10. Kusseling, Francoise S. et al. "Understanding Why Heterosexual Adults Do Not Practice Safer Sex: A Comparison of Two Samples." AIDS Education and Prevention 8 (June 1996): 247-57.

11. Kitahata, M.M. et. al. "Physician Experience and Survival Among Patients with AIDS." Paper presented at the 3rd Conference on Retroviruses and Opportunistic Infections, Washington, D.C., January 28-February 1, 1996.

12. Corea, Gena. The Invisible Epidemic: The Story of Women and AIDS. New York: HarperCollins, 1992.

13. McGinley, Laurie. "States, in Midst of Cash Crunch, Restrict AIDS-Drugs Programs, Report Finds." The Wall Street Journal. July 11, 1997.

14. Kurth, Ann. "An Overview of Women and HIV Disease." In Until the Cure: Caring for Women with HIV. New Haven CT: Yale University Press, 1993.

15. Nicolosi, Alfredo et al. "The Efficiency of Male-to-Female and Female-to-Male Sexual Transmission of the Human Immunodeficiency Virus: A Study of 730 Stable Couples." Epidemiology 5 (November 1994): 570-5.

16. Centers for Disease Control and Prevention. "Recommendations of the U.S. Public Health Service Task Force on the Use of Zidovudine to Reduce Perinatal Transmission of Human Immunodeficiency Virus." MMWR: Morbidity and Mortality Weekly Report 43 (August 5, 1994): 1-20.

17. Osborne, June E. "Public Health and the Politics of AIDS Prevention." Daedalus: Journal of the American Academy of Arts and Sciences 118 (summer 1989): 123-44.

18. Dolcini, M. Margaret et al. "Cognitive and Emotional Assessments of Perceived Risk for HIV Among Unmarried Heterosexuals." AIDS Education and Prevention 8 (August 1996): 294-307.

HIV-Related Internet News
and Discussion Groups
as Professional and Social Support Tools

Sarah C. Fogel

SUMMARY. Information about HIV and AIDS is changing rapidly. HIV is a chronic disease that requires medical, social, and personal management. As of December 1996, more than 293,433 Americans are HIV-positive or have AIDS. HIV has crossed all socioeconomic, gender, and racial barriers, and continues to spread daily. The Internet holds hundreds of HIV-related sites, several of which are available as news and discussion groups. These sites may be used as sources of support or knowledge for people who are HIV-positive, affected by HIV, or for professionals that provide care. *[Article copies available for a fee from The Haworth Document Delivery Service: 1-800-342-9678. E-mail address: getinfo@haworthpressinc.com]*

INTRODUCTION

As of December 1996, more than 293,433 Americans were identified as being HIV-positive or having AIDS.[1] This number represents the entire United States for people diagnosed with AIDS, but only 26 states report HIV infection. It is, therefore, estimated that this number of lives is a gross under-representation of people living with HIV.

Sarah C. Fogel, MSN, RN, ACRN (sarah.fogel@mcmail.vanderbilt.edu), is Assistant Professor of the Practice of Nursing at Vanderbilt University School of Nursing, Nashville, TN.

[Haworth co-indexing entry note]: "HIV-Related Internet News and Discussion Groups as Professional and Social Support Tools." Fogel, Sarah C. Co-published simultaneously in *Health Care on the Internet* (The Haworth Press, Inc.) Vol. 2, No. 2/3, 1998, pp. 79-90; and: *HIV/AIDS Internet Information Sources and Resources* (ed: Jeffrey T. Huber) The Haworth Press, Inc., 1998, pp. 79-90; and: *HIV/AIDS Internet Information Sources and Resources* (ed: Jeffrey T. Huber) Harrington Park Press, an imprint of The Haworth Press, Inc., 1998, pp. 79-90. Single or multiple copies of this article are available for a fee from The Haworth Document Delivery Service [1-800-342-9678, 9:00 a.m. - 5:00 p.m. (EST). E-mail address: getinfo@haworthpressinc.com].

Recent advances in treatment have evolved so that HIV may now be conceptualized as a chronic disease.[2] In the United States and other developed countries, treatment now includes the implementation of a two- to three-drug combination antiretroviral therapy. This combination therapy has successfully slowed the progression of the virus and extended the asymptomatic period of the disease process to an undetermined length of time for many HIV-positive people. Those who can obtain and adhere to the new therapies are able to enjoy an extended healthy life span while living with a chronic disease. However, associated stigma and need for knowledge has not diminished and continues to present a challenge for people living with HIV and professional care-providers.

The purpose of this article is to identify and facilitate access to HIV-related Internet sites that may be used as tools to provide information and social support for HIV-positive people, professional care-providers, and others who are affected by HIV. The concepts of chronic disease, quality of life, and social support will be described in this article to provide background for the understanding of the usefulness of computer-aided support. Specific Internet sites will be identified and a synopsis of their intent will be included.

Background

Chronic disease is a long-term physical or mental impairment that potentially impacts every aspect of life for the person living with the disease. It is a process that requires personal and professional management and adaptation.[3] Chronic disease differs from chronic illness in that the latter involves symptoms such as pain, suffering, or distress and is a state of being sick. People living with chronic disease may be symptom-free for extended periods and consider themselves healthy during those times. This distinction is vitally important in the consideration of self-esteem and associated stigmata that are frequently involved in chronic disease management.

Quality of life is a multidimensional concept that encapsulates virtually every aspect of one's life into a single concept. Quality of life has been defined by Ferrans and Powers as "A person's sense of well-being that stems from satisfaction or dissatisfaction with the areas of life that are important to him/her."[4] Important facets of this definition are the recognition of the individual and the individualized meaning of importance and satisfaction. This definition allows different interpretations of good and useful supports or qualities that are to be determined by the person whose life is at the center of discussion. The definition also infers a temporal quality to quality of life that is fluid or dynamic. In this respect, quality of life is an inextricable concern with chronic disease. Adaptation and management of the chronic disease (that occurs over time) will ultimately impact the perception of quality of life for people living with the disease.

Another concept linked with quality of life is social support. Social support is also a multidimensional concept. Defining attributes of social support emerged from research findings in medicine, nursing, epidemiology, sociology, anthropology, psychology, and psychiatry. They are: satisfying social interaction, affiliation needs, reciprocity, sharing, assistance, understanding, encouragement, reinforcement of self-value/esteem, availability, and provision of needed resources.[5-15] It is important to have several different sources of social support available to people who are HIV-positive because many times more traditional types of supports, such as family, are not readily available.[16-17] The relationship of social support to quality of life has been positively correlated within nursing research.[18] Attempts have been made to establish the same link between health and social support, but much more research and clearer conceptual and operational definitions must first be accomplished.[19]

Several Internet sites that will be discussed in this article provide an opportunity for people who are HIV-positive, or otherwise affected by HIV, to incorporate the defining attributes of social support into their lives. The addition of these attributes (or dimensions) of social support may enhance the person's ability to manage their chronic disease and improve their quality of life.

The other Internet sites included offer news and information about HIV treatment, research, coping with chronicity, and policy. It is a well-established fact that experience and knowledge must work in synchrony to provide appropriate care to people infected with HIV.[20] It is also documented that physicians experienced with HIV/AIDS improve the chances of patient survival and outcomes.[21] The Internet news and discussion groups provide a forum for experienced and less experienced care providers to interact. A recent study reports that the preferred mode of seeking information by American and Canadian physicians is through journals and books, followed closely by consultation with colleagues.[22] The format of electronic discussion groups facilitates this process. It also provides an opportunity for professionals to keep abreast of changing assessment and treatment modalities while providing primary care to people living with HIV disease. Printed publications take approximately 12 to 24 months to begin circulation. In the management of HIV disease, the state of the science may change exponentially within that time.

There is a large part of the community of people living with HIV that is extremely computer literate and active in the pursuit of electronic information transfer and communication. This is exemplified through the diversity of the Internet discussion groups that currently serve virtually every HIV-affected population at this point. One of the problems that will be discussed further in this article is the barrier to electronic access among other HIV-positive populations.

Rationale

Since the late 1980s there has been a growing trend to augment patient knowledge and facilitate communication through the use of computers outside of the hospital. Several projects serve to exemplify this trend within the HIV/AIDS arena. ComputerLink, a project begun in 1988 in Cleveland, served to connect homebound patients living with HIV and nurses.[23] The connection provided information, decision supports, and communication.[24] ComputerLink demonstrated that patients living with HIV, when given the opportunity, are willing to use computers to access information and communicate with one another as well as health care providers. The project also revealed that "computer networks can enhance nursing's ability to intervene with clients experiencing clinical problems."[25]

A similar project, AIDSNET, was begun in 1990 by the Decker School of Nursing at the State University of New York at Binghamton.[26] AIDS-NET provided a small group of homebound people living with HIV access to a nurse consultant with specialized training in AIDS, communication with each other, AIDS newsletters, and research results from around the world.

In 1993, results of piloting the Comprehensive Health Enhancement Support System (CHESS) on people living with HIV were reported.[27] In a six-month period, 39 HIV-positive people used the system 5,520 times for discussion groups, general questions and answers, expert replies to confidential questions, and library and other services. CHESS is being developed to offer a variety of support services to many different populations in health-related crises.

Vanderbilt University Medical Center's Eskind Biomedical Library began an HIV/AIDS outreach project, in 1995.[28] One of the goals for this project was to provide a format for access to electronic HIV-related information that was organized in a way that would facilitate ease of use. The project, HIV/AIDS Community Information Service (http://www.mc.vanderbilt.edu/aidscis/), also provides an interactive component that gives people an online opportunity for obtaining answers to HIV-related questions. Questions continue to come in at a steady rate, which demonstrates continued use of the system three years after its inception.[29]

DESCRIPTION OF SITES

Although the four electronic information systems projects previously discussed are not news or discussion sites, they offer rationale for computer assistance in social support and information retrieval for persons affected by HIV disease. This section lists and briefly discusses samples of HIV/AIDS electronic news and discussion groups. Subscription information is provided. The sites were chosen to represent a wide variety of concern and information for people affected by the HIV pandemic.

News Groups

AIDS

This site provides a forum for open discussion and presents treatment news and information posted on the AIDS Daily Summary. It is a mailing list and gateway of the news group sci.med.aids. To subscribe, send the message *subscribe aids* to majordomo@wubois.wustl.edu. Both basic and advanced level information are presented. This site may benefit anyone working with or affected by HIV/AIDS.

AIDS Education Global Information System (AEGIS)

This electronic resource provides not only a mailing list with discussion opportunities, but also a Web page (http://www.aegis.com/) of basic HIV information that accommodates many different levels of knowledge and needs. Information available from this site comes from many different sources including BusinessWire, PRnewswire, Reuters, and the AIDS Daily Summary. To subscribe to the mailing list, address an e-mail message to aids-request@aegis.com leave the subject field blank and simply include the word *subscribe* as the text of message.

AIDS-News-UK

This resource originates in the United Kingdom and provides the same basic services as the sci.med.aids, including a discussion group. The weekly summaries of HIV and AIDS stories are from the press in the United Kingdom. It allows the participant to obtain a slightly different perspective on HIV-related news information. To subscribe to the mailing list, address an e-mail message to listserv@posnet.co.uk, leave the subject field blank, then type the phrase *subscribe AIDS-News-UK* as the text of the message.

CATIE-News

CATIE-News is the Canadian version of sci.med.aids. Specifically, it is a mailing list for the Community AIDS Treatment Information Exchange (CATIE) and provides news on the latest developments in HIV/AIDS research, treatment, and Canadian policy related to HIV, as well as a forum for discussion. To subscribe to the mailing list, address an e-mail message to masier@catie.ca, leave the subject field blank, then type *subscribe catie-news* as the text of the message.

Journal-Scan

Journal-Scan is a part of CATIE. It provides a weekly distribution of abstracts from articles on HIV/AIDS treatments. There is not an opportunity for discussion at this site, but the value of current journal abstracts is tremendous to people who are HIV-positive, health care providers and others affected by HIV/AIDS. To subscribe, address an e-mail message to masier@catie.ca, leave the subject field blank, then type *subscribe journal-scan* as the text of the message.

Discussion Groups

AIDSACT

AIDSACT is a discussion list for people who wish to communicate with others who are interested or involved in AIDS as a political crisis. The discussion is moderated (as are all of the discussion lists in this article with the exception of the chat room). To subscribe to the discussion list, address an e-mail message to listproc@critpath.org, leave the subject field blank, then type *subscribe AIDSACT* followed immediately by your first name and last name.

CARINGPARENTS

This is a discussion list for parents or any adult who wishes to communicate with other people about how children cope with serious illness. It is open to any adult who is concerned about a child, not only their own, and not only HIV-related. To subscribe, address an e-mail message to listserv@sjuvm.stjohns.edu, leave the subject line blank, then type *subscribe CaringParents* followed immediately by your first name and last name.

Caregivers of People with AIDS

This site is a forum for discussion for anyone who is providing care for someone living with HIV/AIDS. It is a supportive tool for caregivers at any level: a family member, a health care provider, or friend. To subscribe, address an e-mail message to caregivers@sibyllineofbooks.com, leave the subject field blank, then type *subscribe caregivers* followed immediately by your first name and last name.

CRIX-LIST

CRIX-LIST is a discussion group for people who are taking, anticipating taking, or interested in the protease inhibitor, Crixivan. To subscribe or retrieve information files about Crixivan, address an e-mail message to Crix-List@PinkPage.com, then type *subscribe* in the subject field of the message.

Gay Poz

Gay Poz is a discussion group that provides both emotional support and a valuable sharing of medical information for gay men who are living with HIV or AIDS. Participation requires that you answer a few questions before subscribing to the list. To subscribe, address an e-mail message to gaypoz-approval@web-depot.com, then enter a request to join in the body of the message. Once the answers are returned, you will be subscribed and may post messages either with your identification posted or anonymously.

HEMOfact

This is a discussion list for people who are affected by hemophilia. To subscribe, address an e-mail message to listserv@sjuvm.stjohns.edu, leave the subject field blank, then type *subscribe hemofact* followed immediately by your first name and last name as the text of the message.

HIV-DOCS

HIV-DOCS offers a discussion for physicians and other medical professionals such as nurse practitioners who are involved in the care of people living with HIV or AIDS. To subscribe to the discussion group, address an e-mail message to lists@web-depot.com, leave the subject field blank, then type *subscribe hiv-docs* as the text of the message. A second option is available that offers a digest version, which is a condensed version of the day's postings. To subscribe to the digest version only, address a message to lists@web-depot.com, leave the subject field blank, then type *subscribe hiv-docs-digest* as the text of the message.

HIV-LAW

HIV-LAW provides a forum for the discussion of legal and advocacy issues surrounding HIV/AIDS. Both organizations and individuals are welcome to join. Organizations are requested to specify a contact person. To subscribe, address an e-mail message to HIV-law-approval@web-depot.com. In the body of your message, explain who you are, the nature of the services that you provide, and your location. Once subscribed, you may post messages either with identification or anonymously.

HIV-SUPPORT

This site offers two options, either full participation in the discussion group or a digest version. The goal of the discussion list is to provide both emotional support and sharing of medical information for people living with HIV or AIDS. To subscribe to the discussion list, address an e-mail message to lists@web-depot.com, leave the subject field blank, then type *subscribe hiv-support* as the text of the message. To subscribe to the digest version, type *subscribe hiv-support-digest* as the text of a message addressed to lists@web-depot.com. This site offers two options for posting: identified or anonymous.

HOSPICE

Hospice offers a discussion opportunity for people who provide volunteer services to hospices. It is not HIV specific. To subscribe, address an e-mail message to listserv@whitman.edu, then type *subscribe hospice* followed immediately by your first name and last name as the text of the message.

PAIN-L

The usefulness of this site comes from interaction with people using medications for chronic pain that are not normally listed in the HIV-drug literature. The primary topics of this discussion list include health and healing and chronic pain. It is not HIV specific. To subscribe, address an e-mail message to listserv@sjuvm.stjohns.edu, leave the subject field blank, then type *subscribe pain-l* followed immediately by your first name and last name as the text of the message.

SEA-AIDS

This discussion list serves people interested in shaping the response to HIV and AIDS in South East Asia. To subscribe, address an e-mail message to majordomo@lists.inet.co.th, leave the subject field blank, type *subscribe sea-aids* as the text of the message.

Treatments

Treatments is a discussion list for people with personal experience with AIDS-related and HIV-related treatments. Both a "treatments" format and a "treatments-digest" format are available. To subscribe, address an e-mail message to ben@aidsinfobbs.org containing your choice of a list (treatments or treatments-digest) along with your full name.

Marty's HIV/AIDS Support Group and Mailing List (Chat room)

This is an unmoderated forum for discussion that takes place in real time. Unlike all of the other lists mentioned in this article, this site offers unrestricted conversation. Nothing is screened out. It is, however, a restricted site for people who are HIV-positive. This site has been up and running for more than six years, remains very confidential, and requires that you agree to a list of assumptions and guidelines for discussion content. The subscription information and guidelines can be found at its Web page (http://www.smartlink.net/~martinjh/chatnews.htm). To subscribe, address an e-mail message to martinjh@smartlink.net, type *support-group* in the subject field, and include a real name (first and last) along with a clear statement of being HIV-positive (not living with HIV, which may imply affected as opposed to infected) as the text of the message.

BARRIERS TO ACCESS

Many different barriers are present in the assessment of access and use of electronic HIV-related Internet sites. The most obvious is lack of financial resources that prohibits ownership or access to a computer setup or Internet connection. Many cities have public libraries with free access to these types of facilities. Some communities are beginning to set up workstations in social-service organizations or clinics that encourage searching the Net for answers to medical questions.[28] Still, a large percentage of people who are HIV-positive or affected by the pandemic do not have access to these resources.

Another problem or barrier is population specific. HIV is not population specific and affects virtually all communities with no regard to culture or custom. People who are living on the street, injecting drug users, women with children at home, people living in rural areas with restricted telecommunication capabilities, and older people are only a sampling of the populations that may have barriers to access of computer-based technology.

A final barrier is either functional or computer illiteracy. Many adults in the United States, and throughout the world, live their lives without the ability to read. This barrier extends not only to the use of electronic information systems, but to many sources of information. Electronic information systems provide the potential for reducing this barrier through Windows technology and picture presentations. Computer illiteracy is decreasing daily for children through the addition of computer-based learning in most public schools, at least in the United States, but many adults remain undereducated.

CONCLUSION

The previously mentioned barriers all have solutions. The continuation of projects such as ComputerLink, AIDSNET, CHESS, and the HIV/AIDS Outreach Project, increasingly affordable computers, advances in telecommunications, and educational endeavors that reduce illiteracy must continue and increase in frequency. For people who are HIV-positive, it has been demonstrated that computer-assisted learning and Internet-based social support networks are effective.[23-25]

Major concepts that provide rationale for the importance of HIV-related Internet news and discussion groups as professional and social support tools are chronic disease, quality of life, and social support. The listing of news and discussion groups is not exhaustive or complete. It provides a sampling of several sites that many populations may find helpful in the management of personal, medical, or social sequelae attributable to the HIV pandemic.

REFERENCES

1. Centers for Disease Control and Prevention. *HIV/AIDS Surveillance Report* 8 no. 2 (1996): 35.

2. Antoni, M.H. "Cognitive-Behavioral Intervention for Persons with HIV." In *Group Therapy for Medically Ill Patients*. New York: The Guilford Press, 1997.

3. Fogel, S.C. "Development of a Chronic Disease Model for HIV/AIDS Nursing." Presentation at the Tennessee Nurses Association, Nashville, TN, November, 1996.

4. Ferrans, C.E., and Powers, M.J, "Psychometric Assessment of the Quality of Life Index." *Research in Nursing and Health*. 15 (February 1992): 29-38.

5. Antonovsky, A. *Health, Stress, and Coping*. San Francisco: Jossey-Bass, 1979.

6. Boyce, W.T.; Jensen, E.W.; James, S.A.; Peacock, J.L. "The Family Routines Inventory: Theoretical Origins." *Social Science and Medicine*. 17 no. 4 (1983): 193-200.

7. Cobb, S. "Social Support as a Moderator of Life Stress." *Psychosomatic Medicine*. 38 (September-October 1976): 300-14.

8. Cohen, S. "Social Support Systems and Physical Health: Symptoms, Health Behaviors, and Infectious Disease." In *Life-Span Developmental Psychology: Perspectives on Stress and Coping*. Hillsdale, NJ: Lawrence Erlbaum, 1991.

9. House, J.S. *Work, Stress, and Social Support*. Reading, MA: Addison-Wesley, 1981.

10. House, J.S. et al. "Social Relationships and Health." *Science* 241 (July 29, 1988): 540-5.

11. Kahn, R.L., and Antonucci, T.C. "Convoys Over the Life Course: Attachment, Roles, and Social Support." In *Life Span Development and Behavior*. New York: Academic Press, 1981.

12. Kane, C. "Family Social Support: Toward a Conceptual Model." *Advances in Nursing Science* 10 (January 1988): 18-24.

13. Norbeck, J.S. "Social Support." *Annual Review of Nursing Research* 6 (1988): 85-109.

14. Thoits, P.A. "Conceptual, Methodological, and Theoretical Problems in Studying Social Support and Coping as a Buffer Against Life Stress." *Journal of Health & Social Behavior*. 23 (June 1982):145-59.

15. Wolf, T.M. et al. "Relationship of Coping Style to Affective State and Perceived Social Support in Asymptomatic and Symptomatic HIV-Infected Persons: Implications for Clinical Management." *Journal of Clinical Psychiatry*. 52 (April 1991): 171-3.

16. Smith, M.Y., and Rapkin, B.D. "Social Support and Barriers to Family Involvement in Caregiving for Persons with AIDS: Implications for Patient Education." *Patient Education and Counseling*. 27 (January 1996): 85-94.

17. Turner, H.A.; Hays, R.B.; and Coates, T.J. "Determinants of Social Support Among Gay Men: The Context of AIDS." *Journal of Health and Social Behavior* 34 (March 1993): 37-53.

18. Nunes, J.A. et al. "Social Support, Quality of Life, Immune Function, and Health in Persons Living with HIV." *Journal of Holistic Nursing*. 13 (June 1995): 174-98.

19. Green, G. "Social Support and HIV." *AIDS Care* 5 (1993): 87-104.

20. Curtis, J.R. et al. "Physicians' Ability to Provide Initial Primary Care to an HIV-Infected Patient." *Archives of Internal Medicine* 155 (1995): 1613-8.

21. Kitahata, M.M. et al. "Physicians' Experience with the Acquired Immunodeficiency Syndrome as a Factor in Patients' Survival." *New England Journal of Medicine* 334 (March 14, 1996): 701-6.

22. Haug, J.D. "Physicians' Preferences for Information Sources: A Meta-Analytic Study." *Bulletin of the Medical Library Association* 85 (July 1997): 223-32.

23. Simpson, R.L. "Computer Networks Show Great Promise in Supporting AIDS Patients." *Nursing Administration* 18 (Winter 1994): 92-5.

24. Brennan, P.F. "ComputerLink A Computerized Nursing Care Delivery System." *Western Journal of Nursing Research* 14 (April 1992): 239-40.

25. Ripich, S.; Moore, S.M.; and Brennan, P.F. "A New Nursing Medium: Computer Networks for Group Intervention." *Journal of Psychosocial Nursing & Mental Health Services* 30 (July, 1992): 15-20.

26. "AIDSNET, a Program Developed by the Decker School of Nursing." *Computers in Nursing.* 10 (July-August 1992): 145-6.

27. Gustafson, D.H. et al. "CHESS: A Computer-Based System for Providing Information, Referrals, Decision Support and Social Support To People Facing Medical and Other Health-Related Crises." *Proceedings of the Annual Symposium on Computer Applications in Medical Care* (1993): 161-5.

28. Huber, J.T., and Giuse, N.B. "HIV/AIDS Electronic Information Resources: A Profile of Potential Users." *Bulletin of the Medical Library Association.* 84 (October 1996): 579-81.

29. Huber, J.T. Personal Communication. Nashville, TN (August, 1997).

HIV/AIDS Information Resources and Services from the National Institutes of Health

Gale A. Dutcher

SUMMARY. As part of its efforts to combat HIV/AIDS, the Federal government has developed numerous information resources and services. The National Institutes of Health, a leader in biomedical research, has taken a leadership role in the provision of important information services as well. Use of the Internet has facilitated information dissemination and enabled researchers, health care providers, patients, and the general public to have access to the most up-to-date information. *[Article copies available for a fee from The Haworth Document Delivery Service: 1-800-342-9678. E-mail address: getinfo@haworthpressinc.com]*

Information, whether on paper or in electronic form, is crucial in the fight against AIDS. As scientists continue to uncover the secrets of the human immunodeficiency virus, health care providers, patients, and the general public must have access to the most up-to-date information to implement these findings. The Federal government is one of the organizations which must work to ensure that mechanisms for disseminating this information meet these needs.

Gale Dutcher (dutcher@nlm.nih.gov) is currently Special Assistant to the Associate Director for Specialized Information Services (SIS) at the National Library of Medicine, Division of Specialized Information Services, Bethesda, MD.

[Haworth co-indexing entry note]: "HIV/AIDS Information Resources and Services from the National Institutes of Health." Dutcher, Gale A. Co-published simultaneously in *Health Care on the Internet* (The Haworth Press, Inc.) Vol. 2, No. 2/3, 1998, pp. 91-98; and: *HIV/AIDS Internet Information Sources and Resources* (ed: Jeffrey T. Huber) The Haworth Press, Inc., 1998, pp. 91-98; and: *HIV/AIDS Internet Information Sources and Resources* (ed: Jeffrey T. Huber) Harrington Park Press, an imprint of The Haworth Press, Inc., 1998, pp. 91-98. Single or multiple copies of this article are available for a fee from The Haworth Document Delivery Service [1-800-342-9678, 9:00 a.m. - 5:00 p.m. (EST). E-mail address: getinfo@haworthpressinc.com].

The National Institutes of Health (NIH), the federal agency responsible for conducting and supporting research that will lead to better health for everyone, is also responsible for disseminating information to support research, treatment, and prevention related to HIV and AIDS. Progress in these areas depends upon the transfer of information to researchers, health care providers, those who provide HIV-related services, and HIV-infected individuals and their advocates. These audiences have varying needs for information that is critical in the fight against HIV/AIDS. To address these needs, the NIH has ongoing programs as well as new initiatives in several arenas.

The passage of the Health Omnibus Programs Extension Act of 1988 (HOPE, PL100-607) marked the beginning of NIH's involvement in large-scale provision of HIV/AIDS information. This legislation specifically mandated the development of several AIDS-related information services. At that time use of the Internet was not widespread outside of university research settings. However, many of the information services developed initially have since evolved to include an Internet-accessible component.

NIH, like the rest of the Federal government, is committed to making more information available through the World-Wide Web. The NIH home page (http://www.nih.gov) provides organizational links to all of the institutes and program areas, as well as links to important information about grants (http://www.nih.gov/grants/), including the all-important grant application form (http://www.nih.gov/grants/funding/phs398/phs398.html) and instructions. In addition to overarching information such as grants and contracts, the NIH home page provides quick access to HIV/AIDS resources for researchers including the NCI AIDS Malignancy Bank Database (http://cancernet.nci.nih.gov/amb/amb.html) and the NIH AIDS Reagent Program (http://www.niaid.nih.gov/reagent/default.htm). The AIDS Malignancy Bank Database is a collection of tissues and biological fluids with associated clinical and follow-up data from patients with HIV-related malignancies. The specimens and clinical data are available for research studies, particularly those that translate basic research findings to clinical application. The AIDS Reagent Program is a resource of over 1,000 state-of-the art reagents for AIDS research.

Two HIV/AIDS-related services which are accessible from the NIH home page have been developed by consortia of Public Health Service agencies. They have both electronic (Internet) access and voice (telephone) access. These multi-media approaches have been important in a situation where information is changing extremely rapidly and where social factors combine with scientific issues to force the pace of progress and institutional response.

As a result of the HOPE legislation, the National Institute for Allergy and Infectious Diseases (NIAID), the National Library of Medicine (NLM), both components of the NIH, and the U.S. Food and Drug Administration working with the Centers for Disease Control and Prevention developed the AIDS Clinical Trials Information Service (ACTIS–http://www.actis.org). As part of this service, two online databases are available via the Internet from NLM (http://igm.nlm.nih.gov). The AIDSTRIALS database contains detailed descriptions of clinical trials testing different treatments for HIV, AIDS, and related diseases. In addition to a description of the purpose of the trial, sufficient information is included to assist the health care provider or patient in determining eligibility criteria. The AIDSDRUGS database contains information about the agents tested in clinical trials including synonyms and trade names of the drugs. Both databases maintain the records even after the trial has been completed or the drug approved for use.

The ACTIS Web site contains links to information resources that may assist users in understanding clinical trials. For example, online ordering of a video, *HIV/AIDS Clinical Trials: Knowing Your Options*, is available. A database of citations of articles reporting results of clinical trials is also included in this site. These citations are selected from NLM's online databases AIDSLINE and MEDLINE (see below) and are linked to specific trials. Fact sheets and brochures are also available including the basic fact sheet *What is an AIDS Clinical Trial*.

The final component of ACTIS is a toll-free telephone service (1-800-TRIALS-A) for users who prefer to discuss their query with a health information professional, rather than via their computers. This enables virtually any type of user to access the same information regardless of their level of sophistication or access to computers.

The HIV/AIDS Treatment Information Service (ATIS) is a coordinated Public Health Service project sponsored by six Public Health Service agencies: the Centers for Disease Control and Prevention, Agency for Health Care Policy and Research, Health Resources and Services Administration, Indian Health Service, National Institutes of Health, and Substance Abuse and Mental Health Services Administration. The ATIS Web site (http://www.hivatis.org) includes links to NLM's HSTAT (http://text.nlm.nih.gov) database, which supports full-text searching of federally approved treatment guidelines, drafts of new treatment recommendations released by the government, press releases about treatment issues from NIH and other PHS agencies, and an extensive glossary of HIV/AIDS-related terms (http://www.hivatis.org/glossary/) (see Figure 1). ATIS also

FIGURE 1

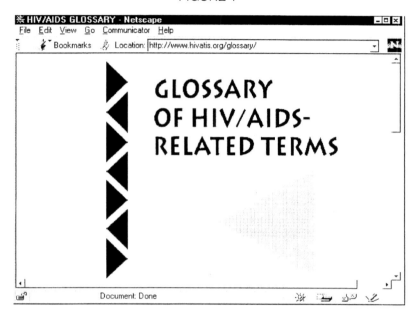

includes a toll-free telephone service (1-800-HIV-0440) to enable those who desire an alternate access method, to obtain this essential information.

In addition to these major collaborative efforts, several of the components of NIH have important HIV/AIDS-related resources available through their Web pages. All of these may be reached either directly or via the NIH home page, which does have a search capability.

The NIH Office of AIDS Research (OAR) (http://www.nih.gov/od/oar/) is responsible for the scientific, budgetary, legislative, and policy elements of the NIH AIDS research program. They plan, coordinate, evaluate, and fund all NIH AIDS research as well as being responsible for the development of an annual comprehensive plan and budget for all NIH AIDS research. This plan (http://www.nih.gov/od/oar/PUBLIC.HTM) focuses on seven areas of emphasis, including information dissemination (http://www.nih.gov/od/oar/OARINFO.HTM).

The Division of Acquired Immunodeficiency Syndrome (DAIDS) (http://www.niaid.nih.gov/research/Daids.htm) of the National Institute of Allergy and Infectious Diseases (http://www.niaid.nih.gov/) is the lead NIH agency for conducting and funding HIV/AIDS-related research.

Their mission is to increase basic knowledge of the pathogenesis, natural history, and transmission of HIV disease and to promote progress in its detection, treatment, and prevention. DAIDS has programs in (1) fundamental basic and clinical research, (2) discovery and development of therapies for HIV infection and its complications, (3) discovery and development of vaccines and other preventive interventions, and (4) training of researchers in these activities. A portfolio of grants and contracts addresses research in these areas. The NIAID Web site includes lists of upcoming scientific meetings, meeting and conference summaries, available research resources such as how to obtain AIDS-related data sets (http://www.niaid.nih.gov/research/research/aidsdataset.htm), links to supported programs such as the AIDS Clinical Trials Group (http://aactg.s-3.com/) and publications such as The Virology Manual for HIV Laboratories (http://www.niaid.nih.gov/daids/vir_manual/), which was developed by NIH and collaborating investigators, and includes detailed consensus virology assay protocols for implementation in multi-site investigations, laboratory biosafety issues, and recommendations for specimen processing, storage, and shipping.

The National Cancer Institute (NCI) also takes a multi-media approach to providing access to information about cancer. The HIV/AIDS-related information included in the Cancer Information Service and CancerNet is the treatment and research about those cancers specifically associated with HIV infection, Kaposi's sarcoma, and certain lymphomas. The Cancer Information Service (1-800-4-CANCER) provides toll-free telephone service for patients and their families, health professionals, and the general public. PDQ is one of NCI's cancer information databases. The database includes up-to-date summaries on treatment, supportive care, screening/prevention, and investigational drugs, in addition to listings of ongoing clinical trials and directories of physicians and organizations involved in cancer care or screening. PDQ may be accessed through the NLM online system (telnet://medlars.nlm.nih.gov) or as part of CancerNet, or from other providers of online databases. The CancerNet (http://cancernet.nci.nih.gov/) service makes the cancer information statements from PDQ and other cancer information from the NCI available quickly and easily via the Internet. Information is available in formats specifically for health professionals or patients and includes care, treatment, screening, and research information. Citations and abstracts from the CANCERLIT bibliographic database are also available.

CANCERLIT (http://www.nlm.nih.gov/pubs/factsheets/online_databases.html) is a bibliographic database covering major cancer topics, including AIDS-related malignancies. In addition to journal articles, the database includes government and technical reports and meeting abstracts. Since 1983, most journal literature has been derived from MEDLINE. CANCERLIT is produced by the National Cancer Institute (NCI) in cooperation with the NLM. Some of the information in CANCERLIT is available through CancerNet as monthly bibliographies on selected topics. The full database may be accessed via Telnet from NLM (telnet://medlars.nlm.nih.gov) or through many other providers of online databases.

The National Center for Research Resources (http://www.ncrr.nih.gov/) is responsible for assisting in the development of the infrastructure for biomedical research. They fund shared clinical, primate, and biotechnology resources for use by investigators supported by all the NIH Institutes and Centers. One such resource is the CRISP database (http://www.ncrr.nih.gov/grants/crisp.htm) which contains information on research ventures supported by the United States Public Health Service (US-PHS). Most of this research falls within the broad category of extramural projects: grants, contracts, and cooperative agreements conducted primarily by investigators at universities, hospitals, and other research institutions. CRISP also contains information on the internal research programs of the NIH and FDA. This Web site also includes lists and descriptions of the research centers, training programs, primate facilities, and biological resources (e.g., repositories and animal models) supported and available to other scientists (http://www.ncrr.nih.gov/access.htm).

The National Library of Medicine (http://www.nlm.nih.gov) serves as a national information resource for research, health care, and the education of health professionals. It is responsible for management and dissemination of biomedical information as well as research and development in information technology. The NLM produces many databases including MEDLINE and specialized subject areas such as HIV/AIDS.

Online databases pre-date the development of the Internet and are still accessible through traditional dial-up services as well as via the Internet. The NIH produces several very important databases containing HIV/AIDS information. In addition to access through NIH services, these databases are licensed by other information providers, universities, and companies.

MEDLINE (http://www.nlm.nih.gov/databases/medline.html), NLM's premier bibliographic database, contains over eight million citations to journal articles from 1965 to the present in the fields of medicine, nursing, dentistry, veterinary medicine, the health care system, and the preclinical sciences. MEDLINE includes citations and abstracts to journal articles about HIV/AIDS and related topics such as virology, immunology, and internal medicine. There are many ways to access MEDLINE, including free via the World-Wide Web (http://www.nlm.nih.gov/databases/freemedl.html).

AIDSLINE (http://www.nlm.nih.gov/pubs/factsheets/aidsline.html), created in 1988, includes citations to literature covering research, clinical aspects, and health policy issues related to AIDS published since 1981. The citations are derived from the MEDLINE, CANCERLIT, and several other NLM databases to create a unified, single source of HIV/AIDS-related references. Additional citations added include meeting abstracts from the International Conferences on AIDS and other AIDS-related meetings, conferences, and symposia. Citations and abstracts from newsletters and special AIDS journals are also included. As with MEDLINE, AIDSLINE is available through many services, including free via Internet Grateful Med (http://igm.nlm.nih.gov).

NLM provides access to HIV/AIDS information produced within and outside the government (http://sis.nlm.nih.gov/aidswww.htm). As a result of a conference held at NLM in 1993, NLM produced a Guide to NIH HIV/AIDS Information Services (http://sis.nlm.nih.gov/aids/index.html). This publication brings together a variety of data about the many HIV/AIDS information-related activities of NIH along with selected PHS offerings (see Figure 2). Links are provided to all the components of NIH and the Public Health Service that have reported HIV/AIDS-related information

FIGURE 2

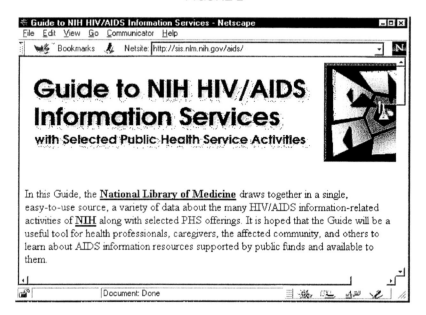

products or services. Included are descriptions of programs and publications as well as links to information available via the Internet. Information resources are categorized by type of information: research, clinical trials, treatment, patient education, professional training, prevention, and general information. Resources are also categorized by format: telephone service, electronic, printed publication, exhibit or educational program. This is a mechanism that provides one-stop shopping to important HIV/AIDS-related information resources from the NIH and Public Health Service.

NLM also provides access to its own products and services. In addition to the AIDSLINE, AIDSTRIALS, and AIDSDRUGS databases (see above), the NLM HIV/AIDS home page (http://sis.nlm.nih.gov/aidswww.htm) provides links to database documentation (http://sis.nlm.nih.gov/aiddoc1.htm), specialized sets of data such as a searchable database of abstracts from the XI International Conference on AIDS (http://sis.nlm.nih.gov/aidsabs.htm), and other publications. A unique resource on this server is the set of chemical structure diagrams for the agents listed in the AIDSDRUGS database (http://sis.nlm.nih.gov/aidsdrg4.htm). This Web page also links to a very selected list of other resources, including those developed by organizations partially funded through NLM's AIDS community outreach activities (http://sis.nlm.nih.gov/aidswwo.htm). NLM has initiated outreach programs to community-based AIDS organizations and patient advocacy groups. The primary purpose of these outreach projects is to enable local groups to develop their information infrastructure and to improve access to information for AIDS patients and the affected community as well as their caregivers. Outreach projects include making community information available through the Internet, providing Internet connections and equipment to community groups, and training in the use of electronic resources.

The National Institutes of Health, along with other agencies of the Public Health Service, has taken a very active role in improving access to important information for those working in the AIDS arena. The information ranges from the highly technical to very basic and is directed to all user constituencies. The goal of these efforts to use the Internet to disseminate information is to enable those who need it to obtain the right information when and where they can best use it.

CDC's HIV/AIDS Resources on the Net

Ann L. Poritzky

SUMMARY. Medical advances are improving the lives of people living with HIV/AIDS, while prevention remains the key to stopping the spread of this devastating disease. The Centers for Disease Control and Prevention (CDC) has created an integrated information delivery system to provide information about HIV/AIDS to the public, health professionals, and people living with HIV/AIDS. Consumers can get information by telephone, mail, and via the Internet. This article provides details about CDC's HIV/AIDS information services and how to access them. *[Article copies available for a fee from The Haworth Document Delivery Service: 1-800-342-9678. E-mail address: getinfo@haworthpressinc.com]*

INTRODUCTION

HIV infection is one of the most significant health challenges facing America today. Although recent medical advances have enabled people with HIV/AIDS to live longer, often with an improved quality of life, it is well known that those who have current information about their illnesses and options fare better than those who do not. Despite promising advances in therapy, prevention still remains the best weapon for stopping the spread of this devastating disease.

Ann L. Poritzky, MBA (aporitzky@aspensys.com), is Senior Communications Specialist with the CDC National AIDS Clearinghouse, Rockville, MD.

[Haworth co-indexing entry note]: "CDC's HIV/AIDS Resources on the Net." Poritzky, Ann L. Co-published simultaneously in *Health Care on the Internet* (The Haworth Press, Inc.) Vol. 2, No. 2/3, 1998, pp. 99-104; and: *HIV/AIDS Internet Information Sources and Resources* (ed: Jeffrey T. Huber) The Haworth Press, Inc., 1998, pp. 99-104; and: *HIV/AIDS Internet Information Sources and Resources* (ed: Jeffrey T. Huber) Harrington Park Press, an imprint of The Haworth Press, Inc., 1998, pp. 99-104. Single or multiple copies of this article are available for a fee from The Haworth Document Delivery Service [1-800-342-9678, 9:00 a.m. - 5:00 p.m. (EST). E-mail address: getinfo@haworthpressinc.com].

In 1987, the Centers for Disease Control and Prevention (CDC) created an integrated information delivery system to provide information about HIV/AIDS to the public and to health professionals. The CDC National AIDS Hotline (1-800-342-AIDS) provides information to the general public on HIV risk and transmission issues and on HIV-antibody testing. The CDC National AIDS Clearinghouse delivers information to those working in the fields of HIV prevention, services, and care.

Since 1987, three additional services have been added to the Clearinghouse:

- The Business and Labor Resource Service, to help businesses and labor unions deal with the challenges of AIDS in the workplace.
- The AIDS Clinical Trials Information Service (ACTIS), a service sponsored by four Public Health Service (PHS) agencies to provide information on federally and privately sponsored clinical trials on HIV/AIDS.
- The HIV/AIDS Treatment Information Service (ATIS), another service sponsored by multiple PHS agencies to provide information about federally approved treatments for HIV/AIDS.

Each of the services has a toll-free telephone number and provides information through the mail and via fax to health care providers and people with HIV infection and AIDS. During the past two years, each service has developed extensive Internet-based services, using the power and speed of e-mail and the Web to share information about lifesaving prevention techniques and treatments with individuals and organizations throughout the world. This article presents an introduction to each information service and its related Internet resources.

CDC NATIONAL AIDS CLEARINGHOUSE (CDC NAC)

The CDC National AIDS Clearinghouse (1-800-458-5231) provides timely, accurate, and relevant information about HIV/AIDS to people and organizations working in prevention, health care, medical research, and support services. The Clearinghouse offers comprehensive reference and referral services; maintains databases on HIV/AIDS; distributes millions of educational materials each year by mail and fax; offers training and technical assistance on finding HIV/AIDS information; and operates two resource centers with extensive collections of scientific and educational materials.

Visit the Clearinghouse's Web pages (http://www.cdcnac.org) to:

- Locate HIV/AIDS testing and counseling centers.
- Find HIV/AIDS-related health and other services in your area.
- Identify specialized materials for prevention, counseling, and training programs.
- Order selected free publications.
- Learn about special services for women; racial and ethnic minority groups; gay men, lesbians, and bisexuals; adolescents and young adults; and other special populations.
- Get information about HIV/AIDS materials in other languages.
- Investigate funding options for community-based programs.
- Read the *AIDS Daily Summary*, a compilation of news articles about HIV/AIDS from major newspapers, journals and news wires around the country.

The site features information about the Clearinghouse's services. Users can download full-text publications on HIV/AIDS, search online databases, preview images of posters and public service announcements, find references to materials in Spanish, and locate numerous links to related sites. An online order form provides an easy way to order free HIV/AIDS materials directly from the Clearinghouse.

The Clearinghouse manages the AIDSNEWS LISTSERV, which sends subscribers the *AIDS Daily Summary*, selected AIDS-related articles from CDC's *Morbidity and Mortality Weekly Report* series, fact sheets, and press releases from various PHS agencies. To subscribe to AIDSNEWS, send an e-mail message to listproc@aspensys.com. Leave the subject line blank. The message line should include: *subscribe aidsnews*, followed immediately by your first name and your last name. A sample e-mail is shown below:

(Sample e-mail request)

To: listproc@aspensys.com

Subject: [leave this blank]

Message: subscribe aidsnews john doe

Another Internet service available from the Clearinghouse is an Anonymous File Transfer Protocol (FTP) site (ftp://ftp.cdcnac.org/pub/cdcnac). The FTP site is particularly useful for people without access to computers with high-level graphic capabilities or fast modems. The large collection of documents available on the FTP site includes the current *HIV/AIDS*

Surveillance Report issued by CDC, CDC NAC guides to AIDS information, and the Clearinghouse's *Standard Search Series* which includes databases on such topics as nutrition, substance abuse, tuberculosis, sexually transmitted diseases, abstinence, and women.

CDC BUSINESS AND LABOR RESOURCE SERVICE (BLRS)

The Business and Labor Resource Service (1-800-458-5231) provides practical assistance, referrals, and materials to managers, supervisors, and labor leaders to help them address specific HIV/AIDS-related issues affecting their employees or organizations. BLRS is part of CDC's *Business Responds to AIDS* (*BRTA*) and *Labor Responds to AIDS* (*LRTA*) Programs. These programs work in partnership with businesses, labor unions, trade associations, public health departments, community-based organizations, and government agencies to promote the development of comprehensive workplace HIV/AIDS programs.

BLRS reference specialists answer questions about setting up employee education programs and preparing HIV/AIDS policies. They also offer guidance to employees who are living with HIV/AIDS and have become too sick to work as well as those whose health has improved and who can return to work.

Visit the Web pages (http://www.brta-lrta.org) to:

- Learn what to do if an employee tells you he or she has the HIV infection.
- Preview a public service announcement that encourages business and labor leaders to get involved in HIV/AIDS workplace prevention initiatives.
- Link to sites with guidelines on safe practices for health care workers and others working with people with HIV infection.

The site includes an overview of CDC's AIDS in the Workplace programs, statements from business and labor leaders about successful initiatives, current information and recent publications on workplace issues, *BRTA/LRTA* public service announcements, and numerous links to related Web sites.

AIDS CLINICAL TRIALS INFORMATION SERVICE (ACTIS)

The AIDS Clinical Trials Information Service (1-800-TRIALS-A) is a toll-free information and referral service that provides details about federally and privately sponsored HIV/AIDS clinical trials. The service is sponsored by CDC, the Food and Drug Administration, the National Institute of Allergy and Infectious Diseases, and the National Library of Medicine.

Visit the ACTIS Web pages (http://www.actis.org) to:

- Get details about new and open clinical trials for treatments and preventive vaccines.
- Search a database of completed clinical trials results.
- Link to related sites.

The site includes information about new clinical trials and provides access to the *AIDS Clinical Trial Results Database*, which includes references to journal articles with interim or final results of HIV/AIDS clinical trials. The database can be searched by key words, as well as by diseases studied, drug names, and types of trials. The references are from the National Library of Medicine's AIDSLINE and MEDLINE databases.

HIV/AIDS TREATMENT INFORMATION SERVICE (ATIS)

The HIV/AIDS Treatment Information Service (1-800-448-0440) offers information on federally approved treatments for HIV infection and related conditions. Using guidelines prepared by federal agencies, the staff of health professionals respond to inquiries about the medical management of HIV/AIDS from people with HIV, their families and friends, and their health providers. ATIS is cosponsored by several agencies, including CDC, the Food and Drug Administration, the Agency for Health Care Policy and Research, the Health Resources and Services Administration, the National Library of Medicine, the National Institute of Allergy and Infectious Diseases, the Substance Abuse and Mental Health Services Administration, and the Indian Health Service.

Visit the ATIS Web Pages (http://www.hivatis.org) to:

- Learn about the latest federally approved HIV/AIDS treatment options.
- Get definitions for terms related to symptoms and treatments.
- Read about protease inhibitors and other approved drugs.
- Find treatment information for women, including treatment during pregnancy.

The ATIS site includes information and publications on HIV/AIDS, a glossary of HIV/AIDS-related terms, and abstracts from selected conferences and meetings.

CONCLUSION

The CDC's information services provide a wealth of reliable HIV/ AIDS information on the Internet for health care consumers. People with all levels of Internet access can get information about research, treatments, workplace issues, community services, health care, and news, along with access to databases and links to other related Web sites. A list of Internet addresses and toll-free telephone numbers for the information services described in this article appears below:

- CDC National AIDS Clearinghouse (http://www.cdcnac.org) 1-800-458-5231
- Business and Labor Resource Service (http://www.brta-lrta.org) 1-800-458-5231
- AIDS Clinical Trials Information Service (ACTIS) (http://www.actis.org) 1-800-TRIALS-A
- HIV/AIDS Treatment Information Service (ATIS) (http://www.hivatis. org) 1-800-448-0440
- File Transfer Protocol Site (ftp://ftp.cdcnac.org/pub/cdcnac)

Complementary
and Alternative Medicine (CAM):
Selected Internet Resources
for HIV/AIDS

Charles B. Wessel
Bruce A. Johnston
Tamera E. Frech

SUMMARY. Although traditional drug treatment options remain a therapeutic mainstay, complementary and alternative medicine continues to entice individuals infected with HIV. This article explores the interaction between alternative medicine and HIV/AIDS as well as the emerging role of the Internet as an information resource. Included is an annotated list of alternative and complementary medicine Internet sites relative to HIV/AIDS. *[Article copies available for a fee from The Haworth Document Delivery Service: 1-800-342-9678. E-mail address: getinfo@haworthpressinc.com]*

Charles B. Wessel, MLS (cbw@med.pitt.edu), is Reference Librarian, Falk Library of the Health Sciences, Health Sciences Library System, University of Pittsburgh, Pittsburgh, PA. He compiles The Alternative Medicine Homepage (http://www.pitt.edu/~cbw/altm.html). Bruce A. Johnston, MLS (baj@city-net.com), is Volunteer Library Supervisor, Pittsburgh AIDS Task Force, Pittsburgh, PA. Tamera E. Frech, BS (tfrech@city-net.com), is Community Relations Associate, Pittsburgh AIDS Task Force, Pittsburgh, PA.

[Haworth co-indexing entry note]: "Complementary and Alternative Medicine (CAM): Selected Internet Resources for HIV/AIDS." Wessel, Charles B., Bruce A. Johnston, and Tamera E. Frech. Co-published simultaneously in *Health Care on the Internet* (The Haworth Press, Inc.) Vol. 2, No. 2/3, 1998, pp. 105-123; and: *HIV/AIDS Internet Information Sources and Resources* (ed: Jeffrey T. Huber) The Haworth Press, Inc., 1998, pp. 105-123; and: *HIV/AIDS Internet Information Sources and Resources* (ed: Jeffrey T. Huber) Harrington Park Press, an imprint of The Haworth Press, Inc., 1998, pp. 105-123. Single or multiple copies of this article are available for a fee from The Haworth Document Delivery Service [1-800-342-9678, 9:00 a.m. - 5:00 p.m. (EST). E-mail address: getinfo@haworthpressinc.com].

105

INTRODUCTION

Traditional medical drug treatment options for HIV/AIDS continue to make headlines on a frequent basis in the United States. Brand names of antiretroviral drugs and protease inhibitors, both approved by the Food and Drug Administration (FDA), as well as those still in the clinical trial phase, are widely recognized throughout HIV/AIDS communities. Multi-drug "cocktail" combinations are now prescribed with increasing frequency. Many individuals with HIV/AIDS in the United States have made remarkable improvements, as evidenced by a recently observed decline in the overall death rate due to AIDS-related illnesses.[1] So why is there such a proliferation of information concerning "alternative and complementary medicine"?

While there are numerous reasons for which individuals may seek non-orthodox methods of HIV/AIDS treatment, a glimpse back to the early history of the AIDS epidemic can provide some insight and perspective. With the recent advent of the protease inhibitors class of drugs and the introduction of viral load testing, it may be difficult to recall the dismal prognosis for people diagnosed with AIDS in the 1980s. The virus was poorly understood; early antiretroviral drugs had intolerable toxic side effects; drug resistance developed quickly; and treatments focused on prevention of debilitating opportunistic infections. During the first decade of the AIDS epidemic, with so few viable options available, many affected individuals began to seriously examine and pursue the possibilities of non-traditional regimens.

Despite significant advances in drug therapy present in 1998, drug combination treatments for HIV/AIDS are not considered to be a cure. Furthermore, when available, these therapies do not provide significant benefit to a large proportion of patients.[2] At best, the present generation of protease inhibitors, in combination with other prophylactic medications, will help transform HIV/AIDS into a manageable, chronic illness, instead of a uniformly debilitating and fatal disease. Both patients and medical professionals not only share valid concerns about the long-term efficacy of these present drugs, but also acknowledge the practicalities of complex drug administration schedules and emerging resistance patterns associated with the continued use of protease inhibitors.

As this new era of AIDS medicine continues to unfold, there is still much that is unknown. Basic research continues to uncover new information about the virus and possible methods to halt its destruction of the immune system. However, there is, as yet, no cure or effective vaccine on the horizon. Complementary and alternative medicine (CAM) may provide some workable options for those individuals who fail to tolerate traditional drug regimens or who seek to reduce potentially unhealthy side effects. An important consideration to recognize in the growth of CAM is the empowerment and sense of independence that such therapies can provide to the consumer. Treatment options, such as guided imagery, acupuncture, and herbal therapy, can engender advantages over traditional medicine: no strict time/dose regimens, free choice of modalities, and control over one's schedule/lifestyle. Lastly, many who choose a type of alternative and/or complementary practice do experience degrees of self-reported success.

Recent efforts have been undertaken not only to more clearly define, categorize, and classify complementary and alternative medicine (CAM), but also to research and clinically document the impact of these various practices on immune system function, pain reduction, and quality of life. This article explores the interaction between alternative medicine and HIV/AIDS. In addition, the emerging role of the Internet as an additional important source of information is examined. Lastly, an annotated directory of Internet sites is described to provide insight and guidance to the wide scope of electronic information resources currently available on alternative and complementary medicine, therapy, and practices.

CAM: TOWARDS A DEFINITION AND CLASSIFICATION

The role of "alternative," "unconventional," "unorthodox," or "complementary" treatments used by people with HIV/AIDS continues to evolve and develop amidst continuing efforts to define and categorize the wide range of practices that comprise this branch of medicine. Ethnocentric considerations, embedded cultural factors, and, most recently, insurance reimbursement issues have contributed to difficulties in acceptable definition and nomenclature. In two landmark articles in 1993 and 1996, Eisenberg and colleagues[3-4] defined unconventional therapies as "medical interventions not taught widely at United States medical schools or generally available at United States hospitals." More recently, Jonas[5] presented to the Advisory Council of the National Institutes of Health (NIH) Office of Alternative Medicine (OAM) a somewhat broader interpretation: "Complementary and alternative medicine is defined through a social process as those practices that do not form part of the dominant system for managing health and disease." In an effort to provide expanded access to the health sciences literature, the Medical Subject Heading (MESH) annotation from the National Library of Medicine was revised in 1996 to define Alternative Medicine as "an unrelated group of unorthodox practices, often with explanatory systems, that do not follow conventional biomedical explanations."[6]

A practical definition of alternative medicine has been presented by Burroughs:

> The term "alternative medicine" is a catch-all phrase which includes therapeutic nutrition, chiropractic, homeopathy, structural and energetic therapies, and mind-body interventions, traditional ethnomedicinal systems such as Chinese medicine, and Ayurveda which combine botanical medicine with other applications, other uses of non-Western and Western herbs, and various treatments which simply have not been accepted by the medical establishment. Alternative medicine has variously been called "natural," "complementary," "holistic," and numerous other terms which refer to elements of a particular modality or tradition.[7]

Burroughs further enumerates that the overall goal of alternative and complementary medical treatments is "to restore strength and balance to weakened systems using a variety of natural modalities: food, herbs and other botanicals, body work, detoxification, etc. . . . , tailored to the individual's specific constitution and condition."[7]

Against the backdrop of these varying definitions, numerous classification schemes have emerged that attempt to address a wide variety of terms, techniques, and practices associated with alternative and complementary medicine. A recent study by Carwein[8] identified six major categories with thirty-three accompanying therapies that comprised commonly used alternative approaches. The six categories and corresponding therapies included: Relaxation (hypnosis, music therapy, relaxation therapy, meditation, biofeedback, and imagery); Touch (massage, reflexology, therapeutic touch, chiropractic therapy, acupressure, acupuncture, Reiki, exercise, and yoga); Diet (macrobiotic diet, Lifestyle, and wheat germ); Ingestion (ozone, hydrogen peroxide, Manchurian mushrooms, folk remedies, megavitamin therapy, homeopathy, aromatherapy, and herbal medicine); Spiritual (Ayurveda, Chakra balancing, spiritual healing); and Self-Help (prayer, self-help groups, laughter/humor, and crystal therapy).

Eisenberg's 1993 analysis of the utilization of unconventional medicine in the United States included eighteen categories of therapy, including relaxation techniques, imagery, megavitamin therapy, biofeedback, massage, and energy healing. A detailed classification was developed and later refined by the Office of Alternative Medicine at the National Institutes of Health based on a multidisciplinary workshop held in 1992. Seven broad "fields of practice" and sixty-three specific therapies were initially identified as pertaining to alternative and complementary medicine (see Table 1).[9]

The Internet-based A-Z Alternative Health Directory (http://www.a-z.

co.uk/health/directory/) provides over seventy categories such as aroma-therapy, chelation therapy, iridology, radiance technique, and zero balancing among its current listings of products, services, and vendors.

In addition, many alternative and complementary therapies support a myriad of specific techniques and practices, each of which has distinct nomenclature and synonyms. For example, an on-going list of techniques, methods, and practices maintained by Massage Therapy Web Central (http://www.qwl.com/mtwc/guide/techniques.html) contains over 140 distinct styles of massage therapy alone, including Amma Therapy, Connective Tissue Massage, Esalan Style Massage, Pfrimmer Deep Muscle Therapy, and Somatoemotional Release.

OVERVIEW OF HIV/AIDS CAM STUDIES

Several types of limited studies have been conducted in recent years to ascertain not only the extent to which HIV/AIDS patients pursue alternative and complementary treatments, but also to evaluate which therapies and modalities are utilized. Carwein[8] analyzed surveys from 127 HIV-positive individuals to report that 20 (16%) of the study population had utilized alternative therapies prior to seroconversion, while all 127 (100%) used alternative medicine after becoming infected. Further analysis revealed 62 (48%) used relaxation methods; 63 (50%) utilized touch therapy; 36 (28%) made use of diet therapies; 35 (27%) practiced spiritual techniques, and 80 (63%) adopted self-help methods. Carwein's data indicated that 76 (60%) respondents claimed that alternative treatments were beneficial to them, with 46 (36%) of the total study population indicating that the chosen alternative therapies helped a great deal.

This study further attempted to delineate the respondents' purposes for utilizing alternative and/or complementary modalities. Few respondents engaged in any specific therapy to treat opportunistic infections (OI) directly. Of the thirty-three available therapies listed on the survey instrument, eighteen were not utilized at all, and the remaining fifteen each were used by only one or two individuals as potential "cures" for opportunistic infections. Three other purposes surveyed for choosing alternative and/or complementary medicine included: increasing quality of life/well-being, helping to control symptoms, and delaying disease progression. In each of these three areas, laughter/humor, relaxation, meditation, music, and exercise modalities each were utilized by at least twenty respondents, with many respondents reporting simultaneous utilization of several therapies.

A study conducted by Anderson et al.[10] in 1993 reported that of 184 HIV-positive and AIDS patients receiving traditional medical care with

TABLE 1. Classification of Complementary/Alternative Medicine (CAM) Practices

Office of Alternative Medicine (OAM), National Institutes of Health (NIH)

Classification	Description
Alternative Systems of Medical Practice	Health care ranging from self-care according to folk principles, to care rendered in an organized health care system based on alternative traditions or practices
Bioelectromagnetic Applications	The study of how living organisms interact with electromagnetic (EM) fields
Diet, Nutrition, Lifestyle Changes	The knowledge of how to prevent illness, maintain health, and reverse the effects of chronic disease through dietary or nutritional intervention
Herbal Medicine	Employing plants and plant products from folk medicine traditions for pharmacological use
Manual Healing	Using touch and manipulation with the hands as a diagnostic and therapeutic tool
Mind/Body Control	Exploring the mind's capacity to affect the body, based on traditional medical systems that make use of the interconnectedness of mind and body
Pharmacological and Biologic Treatments	Drugs and vitamins not yet accepted by mainstream medicine

Modalities

Acupuncture
Antroposophically Extended Medicine
Ayurveda
Community-Based Health Care Practices
Environmental Medicine
Homeopathic Medicine
Latin American Rural Practices

Native American Practices
Natural Products
Naturopathic Medicine
Past Life Therapy
Shamanism
Tibetan Medicine
Traditional Chinese Medicine

Blue Light Treatment and Artificial Lighting
Electroacupuncture
Electromagnetic Fields
Electrostimulation and Neuromagnetic Stimulation Devices
Magnetoresonance Spectroscopy

Changes in Lifestyle
Diet
Gerson Therapy

Macrobiotics
Megavitamins
Nutritional Supplementation

Echinacea (purple coneflower)
Ginger Rhizome
Ginkgo Biloba Extract
Ginseng Root

Wild Chrysanthemum Flower
Witch Hazel
Yellowdock

Acupressure
Alexander Technique
Biofield Therapeutics
Chiropractic Medicine
Feldenkrais Method
Massage Therapy

Osteopathy
Reflexology
Rolfing
Therapeutic Touch
Trager Method
Zone Therapy

Art Therapy
Biofeedback
Counseling
Dance Therapy
Guided Imagery
Humor Therapy
Hypnotherapy

Meditation
Music Therapy
Prayer Therapy
Psychotherapy
Relaxation Techniques
Support Groups
Yoga

Anti-oxidizing Agent
Cell Treatment
Chelation Therapy

Metabolic Therapy
Oxidizing Agents (Ozone,
Hydrogen Peroxide)

antiretroviral drugs (AZT), 74 (40%) also utilized alternative and complementary therapies. Further demographic analysis revealed a correlation between use of alternative and complementary medicine and duration of known seropositivity by the respondent. Forty-nine (52%) of the 95 respondents who had been diagnosed as HIV-positive at least two years prior to the Anderson survey reported utilization of at least one alternative and/or complementary modality. In contrast, 25 (30%) of the 84 recently diagnosed (less than two years before the survey was conducted) individuals had used CAM therapies. In addition to interactions with friends (55%), information about alternative and complementary medicine was obtained from printed newsletters (45%) and books (32%). Support groups provided a significant amount of information for 30% of the respondents.

In a 1994 study at the HIV Outpatient Clinic of the Medical Center of Louisiana at New Orleans, Bates and colleagues[11] utilized a convenience sampling of 287 clients to report that 31% of the respondents used one or more types of alternative and/or complementary medicine. While the median number of therapies used per client was two, additional analysis revealed the following utilization patterns: 80% vitamin/mineral supplements; 37% imagery/meditation; 37% spiritual healing; 37% dietary regimens; 21% herbal therapy; 12% massage; 8% non-prescribed medications; and 10% other types of CAM.

A recent prospective, longitudinal study conducted by Singh and colleagues[12] in 1996 reported that 30% (17/56) of HIV-positive patients in the study population used alternative and/or complementary therapies. Singh concluded that "recourse to nontraditional therapy is common among patients with HIV." After an analysis of psychological characteristics of the limited study population of fifty-six males in a Veterans Administration Medical Center (VAMC) HIV clinic, the study further observed that "patients who choose nontraditional remedies do so not because they are depressed or emotionally disturbed, but rather because they seek greater control of the outcome of their disease."

The consistent results reported by these limited, yet varied studies reveal that the use of alternative and complementary medicine in HIV/AIDS communities is widespread, and correlates closely with recent national data reported by Eisenberg, indicating that one in three persons in the United States has used at least one "unconventional" therapy in the past year. In addition, the types and styles of treatment pursued vary widely, and are affected not only by geographical factors, but also by demographic determinants such as gender, race, age, and education level. Also, these studies in the HIV/AIDS populations indicate that the major reasons for utilizing alternative and complementary therapies cluster around five issues: improving quality of life, enhancing feelings of well-being, increasing self-management/control, slowing disease progression, and ameliorating side effects of antiretroviral drug therapy.[13-14]

CAM AND THE INTERNET

The continual dissemination of information of all types, formats, and levels not only to targeted, at-risk populations, but also to the general public remains a critically important component of all HIV/AIDS treatment, education, and prevention efforts. Not only must relevant information be made available to emerging risk groups such as adolescents, African Americans, bisexuals, women, Hispanics, and older persons, but also existing informational campaigns must be re-formatted and reinforced to those risk groups identified early in the AIDS epidemic, including gay men and intravenous drug users. Siegel, in a recent investigation into access to HIV-related information, care, and services among HIV-positive men from two minority groups, contended that

> it is reasonable to expect that access to disease and treatment-related information about HIV/AIDS can be an important factor in successful psychosocial adaptation to living with the disease. Such information can guide decision making about lifestyle, health behaviors, and therapeutic options, as well as enhance feelings of control over one's life and illness.[15]

As the role and increasing acceptance of alternative and complementary medicine within the HIV/AIDS communities continues to expand, the demand for additional information will also increase. The necessity for current and timely information, coupled with the instantaneous and international aura that envelopes the Internet and World-Wide Web (WWW), has spawned a significant presence of HIV/AIDS CAM-related information on the Internet. A wide spectrum of materials can be located on the Internet: government-produced Web sites, research-based information, academic pathfinders/guides, specialized single modality-based sources, ethnocentric-oriented locations, commercial/retail advertisements, and personal anecdotal home pages.

As with all Internet-produced documents and sources, common sense must be applied in the evaluation of the information presented. In addition to critical appraisal of the source and authority of the Internet document, an evaluation of the timeliness, context, and motive of the document and its producer must also be carefully considered. Tate[16] and Alcorn[17] both published articles that provide practical guidelines for the evaluation of new HIV/AIDS therapies which can readily be applied to the appraisal of HIV/AIDS CAM information residing on the Internet. Both documents can be accessed on the Web (http://www.critpath.org/aric/library/altern05.htm).

The following Internet Web sites have been selected, evaluated, and annotated to present an overview of the alternative and complementary medicine resources available as of September 1997. While several of the sites are general in content, most of the listings are specific to HIV/AIDS resources. In addition to the numerous sites reviewed in this article, many of the postings also contain extensive linkages that will provide further informational resources for exploration.

SELECTED INTERNET SITES

Selected Internet sites are arranged into five categories: sites dedicated specifically to CAM in AIDS/HIV; databases for locating journal articles; general CAM Sites that are comprehensive; modality-specific CAM therapy or treatment; and practitioner's directories for locating specialists in specific CAM modalities.

Since the World-Wide Web changes daily, it should be checked regularly for new sites and updated resources. Information about any of the modalities listed in Table 1 can be located by Internet search engines, including HotBot (http://www.hotbot.com), AltaVista (http://www.altavista.com) or Lycos (http://www.lycos.com). Also, use any of the modalities named in Table 1 to search newsgroups at DejaNews (http://www.dejanews.com).

AIDS/HIV CAM Sites

AIDS Alternative Treatment
(http://www.critpath.org/alt.htm)

This project "was founded by persons with AIDS (PWAs) to provide treatment, resource, and prevention information in wide-ranging levels of detail–for researchers, service providers, treatment activists, but, first and foremost, for other PWAs who often find themselves in urgent need of information quickly and painlessly." The Alternative Treatment index explores alternative, complementary, and unconventional approaches to treatment with the intention of helping PWAs to differentiate rational therapy from quackery. This site includes an extensive index with reports, articles, and information on specific treatment modalities.

AIDS Treatment News
(http://www.immunet.org/immunet/atn.nsf/page/i-latest)

Immunet
(http://www.immunet.org/immunet/atn.nsf/homepage)

AIDS Treatment News is an internationally recognized bi-monthly newsletter which is an excellent resource for persons living with HIV/ AIDS who are looking for information on new therapies. All articles are fully searchable and indexed to provide easy access to CAM information and specific therapies. Recent articles include: Chinese Medicine: Where Does It Work Best in HIV/AIDS? (http://www.immunet.org/immunet/atn.nsf/ page/a-224-01); DHEA: Modest Viral Load Reduction in Patients (http:// www.immunet.org/immunet/atn.nsf/page/a-252-06); Medical Marijuana: Le-gal Issues for Physicians (http://www.immunet.org/immunet/atn.nsf/page/ a-261-09); and Some Vitamins Associated with Decreased Risk of AIDS and Death (http://www.immunet.org/immunet/atn.nsf/page/a-214-07).

Alternative Therapies (and Other Theories)
(http://www.critpath.org/aric/pwarg/links06.htm)

ARIC (AIDS Research Information Center) provides HIV/AIDS medi-cal resources and information primarily to the communities of Baltimore, Maryland and Washington, D.C. The AIDS Link Index goal is to provide access to sites offering quality information on AIDS medical research and treatment. This index is a good starting point for general resources as well as to specific HIV/AIDS CAM therapy information.

Alternative AIDS Therapies (Index)
(http://www.critpath.org/aric/library/altern.htm)

The articles at this site are "intended only as a starting point for inquiry into this fascinating subject. These articles will give you a basic under-standing of the most popular alternative treatments now used by People with HIV/AIDS, and an idea of the role nutrition can play in AIDS medi-cal care." Links are available to the following articles: Alternative AIDS Therapies: An Historical Review; Common Alternative Therapies; Under-standing Vitamins; Herbal Medicines: A Mild Warning; and How to Eval-uate New AIDS Therapies. Common Alternative Therapies provides an overview and excellent description of the compounds as well as references to medical and alternative literature.

Alternative and Complementary Medicine, Positive Living:
A Practical Guide for People with HIV
(http://www.metrokc.gov/health/apu/workbook/chapter2.htm#alternative)

This chapter gives guidance and suggestions for selecting alternative therapists and treatments.

AIDS Treatment Data Network: Alternative Treatments
(http://204.179.124.69/network/altx.html)

At this site, current literature on CAM from Internet resources, newsletters, and scientific and medical journals is presented. This site is updated and revised frequently.

Project Inform: Alternative Treatments
(http://www.projinf.org/hh/alternative.html)

Included at this location are reviews of the following areas: An Alternative Treatment Activist Manifesto; Alternative Therapies for AIDS: An Historical Review; Common Alternative Therapies; Understanding Vitamins; Traditional Chinese Medicine; Antioxidants, Oxidative Stress, and NAC; Evaluating New or "Alternative" Treatments; The Dietary Supplement Health and Education Act of 1993 (S.784/H.R.1709); Kombucha: A Dubious "Cure"; and Alternative Treatment Addenda Sheet.

Bastyr University, AIDS Research Center
(http://www.bastyr.edu/research/recruit.html)

The Center is investigating the forms and patterns of alternative medicine use and effectiveness of treatments. This site contains links to AIDS/HIV resources and the Center's newsletter.

The Body: A Multimedia AIDS/HIV Information Source
(http://www.thebody.com)

This is a comprehensive AIDS/HIV resource with sections covering all aspects of CAM, diet and nutrition, and wellness. The section called Research, and Experimental and Alternative Approaches (http://www.thebody.com/treat/expdrugs.html) provides links to areas such as clinical trials, gene therapy, alternative, and holistic therapies. The Alternative and Holistic Therapies area (http://www.thebody.com/treat/altern.html) is a compilation of links to articles and resources which include AIDS Treatment

News, Natural Health Magazine, PWA Health Group Newsletter, and Women Alive. Articles and resources are indexed into the following categories: General Alternative Medicine; Alternative Drug Treatments; Chinese Medicine; Herbal Remedies; and Fraud. The section on Diet and Nutrition (http://www.thebody.com/dietnut.html), which provides similar types of resources, is indexed into categories of food safety, vitamins, and wasting.

Critical Path AIDS Project: Buyers Clubs
(http://www.critpath.org/docs/buyers.htm)

Buyers Clubs help PWAs obtain substances that are alternative and experimental, and may not be approved or sanctioned by the Food and Drug Administration (FDA). This site contains a list of these clubs.

AIDS Research Information Center, Inc: The Carter AIDS Indexes
(http://www.critpath.org/aric/rtrp/index1.htm)

Originally part of ACT UP, New York's Real Treatments for Real People Project, The Carter AIDS Index provides access to new, experimental, over-the-counter, prescription, FDA-approved non-AIDS medications, and clinical trial treatments. Included are sections on Nutrients/Vitamins as well as Immune-Modulators/Anti-HIV Therapies. These sections are in tabular format with the substance or treatment overview annotated with manufacturer, access, significant descriptive information, and references. This site also contains a PWA resource guide to patient assistance programs, buyers clubs, and links to additional resources.

DAAIR, The Direct AIDS Alternative Information Resources
(http://daair.immunet.org/daair/membinfo.nsf)

This is a members-only not-for-profit buyers club for nutrients, natural therapies, and mind/spirit practices. Contents include treatment information sheets on specific substances, such as Alpha Lipoic Acid, Cat's Claw, Glutamine, etc. These sheets provide extensive referenced information on the various substances. The site contains basic information on food safety, biology of AIDS/HIV, and monitoring of laboratory results. Sections called "treatment buzz" as well as "conference information" provide additional information to members.

The Florida AIDS Health Fraud Task Force
(http://www.applicom.com/tcrs/Fraud.htm)

Task Force membership includes representatives from community, state, and county agencies; AIDS service organizations, buyers clubs; and private business. Their role is to protect against fraudulent activities associated with HIV/AIDS. "AIDS fraud is promotion of an AIDS-related health product, treatment, or service known to be false or labeled with unsupported claims. Fraud can include, but is not limited to: treatment, nutrition, mechanical devices, burial fees, drugs and supplements. Victims of fraud may include not only people with HIV/AIDS, but partners, family and friends as well."

HIV/AIDS Treatment Information Service (ATIS)
(http://www.hivatis.org/)

This site provides information about federally approved treatment guidelines for HIV and AIDS. Access to this site can assist individuals seeking CAM therapies to make informed decisions on appropriate treatments.

Immune Enhancement Project
(http://www.creative.net/~iep/)

The goals of this project are to give treatment information and education on traditional Chinese medicine (TCM) to persons with HIV, chronic fatigue syndrome, Epstein-Barr virus, hepatitis, cancer, and other immune disorders. Care is focused on individual treatments as well as integrating these treatments into Western medicine.

Inter-Q-Zone: An Internet Magazine
for Transgender/Gay/Lesbian/Bisexual/HIV+ People and Their Friends
(http://www.aidsinfonyc.org/Q-zone/index.html)

Addressing issues of health, nutrition, spirituality, and community is the goal of this electronic magazine. This site contains the Nutrition Access Project which includes Nutrition for Life: a Guide to Healthful Living with HIV by Halley Low.

The PWA (People With AIDS) Health Group
(http://www.aidsnyc.org/pwahg/)

"Dedicated to providing access to promising or experimental treatments for HIV/AIDS, and to making more broadly accessible those few treatments already available." Included are informational sheets on substances, buyers club information, and access to their newsletter "Notes from the Underground."

Databases

The National Library of Medicine's MEDLINE and HIV/AIDS databases provide access to journal articles and conference presentations. This service, available over the WWW, is free of charge. For MEDLINE, AIDSLINE, AIDSDRUGS, and AIDSTRIALS access, use Internet Grateful Med (http://igm.nlm.nih.gov/). For MEDLINE use only, consult the PubMed interface (http://www.ncbi/nlm.nih.gov/PubMed/). The National Institutes of Health, Office of Alternative Medicine has an instructional page, Alternative Medicine Research Using MEDLINE (http://altmed. od.nih.gov/oam/what-is-cam/medline.shtml). Traditionally, the content and scope of these databases have been primarily allopathic (traditional Western medicine).

General CAM Sites

Alternative Medicine: Health Care Information Resources for Patients, Their Families, Friends and Health Care Workers
(http://www-hsl.mcmaster.ca/tomflem/altmed.html).

This page is arranged for easy access by modality as well as by medical system. Most sites are American but those which are Canadian are designated by a Canadian Flag. This site has links to health quackery at Fraud Links (http://www-hsl.mcmaster.ca/tomflem/fraud.html).

Bastyr University Library, Internet and On-Line Resources
(Alternative Medicine)
(http://www.halcyon.com/libastyr/netbib.html)

This resource contains a comprehensive index arranged by general indexes, holistic, herbal and botany, nutrition, diseases and conditions, oriental medicine, modalities, electronic mailing lists, and newsgroups. The mailing lists and newsgroups sections provide extensive options for inquiring about HIV treatment information which is modality specific to the individual alternative system of medical practice.

An Index of Alternative Health Sites on the Web
(http://rain-tree.com/links2.htm)

Compiled by the The Raintree Group, this site provides a general index to CAM resources with links to herbs, vitamins, other CAM indexes and newsletters.

Office of Alternative Medicine (OAM),
National Institutes of Health (NIH)
(http://altmed.od.nih.gov/)

The purpose of this NIH office is to "facilitate the evaluation of alternative medical treatment modalities to determine their effectiveness, provide a public information clearinghouse and a research training program. The OAM does not act as a referral center for locating therapies or practitioners. The OAM conducts research."

SageWays: Alternative/Complementary Health Care Directory
for Houston and Beyond
(http://www.sageways.com/start.html)

This is a site which contains general information about CAM and provides a newsletter which covers a variety of topics including HIV/AIDS. Recent articles include: Boxwood Herbal Extract Shows Promise as an HIV Antiviral; Healing Power of Prayer; and HIV Antiviral Effects of Hyperbaric Oxygen Therapy.

United States Food and Drug Administration (FDA)
(http://www.fda.gov)

FDA "is a public health agency, charged with protecting American consumers by enforcing the Federal Food, Drug, and Cosmetic Act and several related public health laws." The FDA is a regulatory agency with responsibility for the safety and effectiveness of foods, drugs, biological products, medical devices and cosmetics. For more information on the FDA's responsibilities, see its Mission and Mandates page (http://www.fda.gov/opacom/7mandate.html). The FDA's Office of Special Health Issues maintains an HIV and AIDS page (http://www.fda.gov/oashi/hiv.html) arranged into several categories including: Status of HIV/AIDS Therapies; HIV Testing; Clinical Trials and Drug Development; and Barrier Products. The section on Evaluating Medical Therapies provides links to sites dealing with health fraud and how to evaluate and choose appropriate medical treatment/therapy.

Modality-Specific

American Botanical Council, Austin, Texas
(http://www.herbalgram.org/)

This Council works to educate the public about the benefits and uses of herbs and plants. It also offers a variety of resources and e-mail contacts for specific herb questions.

Herb Research Foundation
(http://www.herbs.com)

The Foundation works to improve world health and well-being through herbs. Access to the collection is available for a fee, in order to cover research time, copying, and shipping costs. Members receive discounts and priority service.

Institute for Traditional Chinese Medicine:
HIV/AIDS Information Page
(http://www.europa.com/~itm/hiv.htm)

"The Institute for Traditional Medicine (ITM) provides education, conducts research, and offers therapeutic programs with a focus on natural healing techniques, such as herbal formulas, acupuncture, massage, diet, nutrition, and general health care." Examples of content include: Chinese Herbal Therapies for HIV and Spiritual Aspects of Living with HIV. The site also contains natural therapy guides for medical practitioners on vitamins, proteins and fats, antioxidants, Chinese herbs, polypharmacy, and various physical therapies.

The Yoga Group: Yoga for HIV/AIDS
(http://www.yogagroup.org/)

A non-profit organization that provides free Yoga classes to persons living with HIV/AIDS, and assists Yoga instructors to initiate similar programs. Yoga assists with "anxiety reduction, stress management, with a form of exercise that can be adapted to one's level of energy and stamina."

Practitioners' Directories

Acupuncture.com
(http://www.Acupuncture.com/)

This site has extensive resources on acupuncture as well as alternative medicine resources. It is a good starting point for acupuncture information, including state licensure laws and insurance companies that cover acupuncture. Included is Acupuncturists in Your Area (http://www. Acupuncture.com/Referrals/Ref.htm), a list of practitioners arranged by country and state.

American Academy of Medical Acupuncture
(http://www.medicalacupuncture.org/)

The goal of this academy is "to promote the integration of concepts from traditional and modern forms of acupuncture with Western medical training and thereby synthesize a more comprehensive approach to health care." A referral list of practitioners (http://www.medicalacupuncture.org/referral.html) is available arranged by state.

American Association of Naturopathic Physicians
(http://www.infinite.org/Naturopathic.Physician/)

This association represents licensed naturopathic physicians in the United States. Twelve states and the District of Columbia license Naturopathy. A directory of naturopathic physicians by state is included.

Federation of Chiropractic Licensing Boards
(http://www.fclb.org/fclb/)

The Federation works to promote standards and cooperation among chiropractic licensing boards. In addition, it provides information on state licensure.

National Center for Homeopathy
(http://www.homeopathic.org/)

This Center provides education to the public on homeopathy and resources for homeopaths. This site provides a searchable practitioners directory (http://www.homeopathic.org/nchsearch.htm) accessible by zip code, name, or state. The directory contains practitioners that describe their practice as being at least 25% homeopathic. It is not a directory for evaluating a practitioner's credentials, competence, or homeopathic knowledge.

National Certification Board for Therapeutic Massage and Bodywork
(http://www.ncbtmb.com/)

The NCBTMB certifies massage therapists and bodyworkers. Certification recognizes practitioners who meet "standards of proficiency and who make a commitment to the profession by upholding high practical and ethical standards." This site has a practitioner's directory of certified therapists (http://www.ncbtmb.com/ncb-database/query.htm) which can be searched by zip code.

REFERENCES

1. Centers for Disease Control and Prevention (CDC). "Update: Trends in AIDS Incidence-United States, 1996." *Mortality and Morbidity Weekly Report* 46 (September 19, 1997): 861.

2. "Setbacks For Many On Drugs For AIDS." *New York Times* (September 30, 1997): F4.

3. Eisenberg, David M.; Kessler, Ronald C.; Foster, Cindy et al. "Unconventional Medicine in the United States: Prevalence, Costs, and Patterns of Use." *New England Journal of Medicine* 328 (January 28, 1993): 246-252.

4. Eisenberg, David M. "Advising Patients Who Seek Alternative Medical Therapies." *Annals of Internal Medicine* 127 (July 1, 1997): 61-69.

5. Jonas, William. "Dr. Jonas Addresses Advisory Council-February, 1996." *Complementary and Alternative Medicine at the NIH* 3 (1996): 1.

6. Pavek, R. "New MeSH Terms Add Accessibility to Alternative Medicine Literature." *Alternative Therapies in Health and Medicine* 2 (1996): 25-28.

7. Burroughs, Carola. "Alternative AIDS Therapies: An Historical Review (http://www.critpath.org/aric/library/altern01.htm)." Gay Men's Health Crisis (GMHC) Treatment Issues 7 (November, 1993).

8. Carwein, Vicky L., and Sabo, Carolyn E. "The Use of Alternative Therapies for HIV Infection: Implications for Patient Care." *AIDS Patient Care and STDs* 11 (February, 1997): 79-85.

9. National Institutes of Health. Office of Alternative Medicine. *Alternative Medicine: Expanding Medical Horizons*. Washington, DC: U.S. Government Printing Office, 1994.

10. Anderson, Warwick; O'Connor, Bonnie B.; MacGregor, Rob Roy; and Schwartz, J. Sanford. "Patient Use and Assessment of Conventional and Alternative Therapies for HIV Infection and AIDS." *AIDS* 7 (April, 1993): 561-566.

11. Bates, B.R.; Kissinger, P.; and Bessinger, R.E. "Complementary Therapy Use Among HIV-Infected Patients." *AIDS Patient Care and STDs* 10 (February, 1996): 32-36.

12. Singh, Nina; Squire, Cheryl; Sivek, Carla et al. "Determinants of Nontraditional Therapy Use in Patients with HIV Infection." *Archives of Internal Medicine* 156 (January 22, 1996): 197-201.

13. MacIntyre, Richard C.; Holzemer, William L.; and Philippek, Marianna. "Complementary and Alternative Medicine and HIV/AIDS. Part I: Issues and Context." *Journal of the Association of Nurses in AIDS Care* 8 (January-February, 1997): 23-31.

14. MacIntyre, Richard C.; and Holzemer, William L. "Complementary and Alternative Medicine and HIV/AIDS. Part II: Selected Literature Review." *Journal of the Association of Nurses in AIDS Care* 8 (March-April, 1997): 25-38.

15. Siegel, Karolynn; and Raveis, Victoria. "Perceptions of Access to HIV-Related Information, Care, and Services Among Infected Minority Men." *Qualitative Health Research* 7 (February, 1997): 9-31.

16. Tate, Larry; and Delaney, Martin. "Hope, Folly, or Fraud?" *Project Inform Perspectives* (April, 1992).

17. Alcorn, Keith. "Evaluating New Therapies." Critical Path AIDS Project (Summer, 1994).

The Clinical Management of HIV Disease: Internet Resources on the World-Wide Web

J. Michael Howe

SUMMARY. HIV disease has had a significant impact on society and, perhaps more than any other illness, has profoundly affected the way in which medical, legal, public health, and information services are provided. Fortunately, the epidemic arrived at a time of unprecedented strides in biotechnology and information access. This article describes selected World-Wide Web resources about the clinical management of HIV disease. *[Article copies available for a fee from The Haworth Document Delivery Service: 1-800-342-9678. E-mail address: getinfo@haworthpressinc.com]*

INTRODUCTION

HIV disease has had a significant impact on society and perhaps more than any other illness has profoundly affected the way in which medical, legal, public health, and information services are provided. Fortunately, the epidemic arrived at a time of unprecedented strides in biotechnology and information access. Had the first cases of AIDS been reported in the decade of the fifties rather than the eighties, for example, individuals

J. Michael Howe (hivinfo@itsa.ucsf.edu) is Manager/Medical and Biological Sciences Librarian at the VA AIDS Information Center located at the VA Medical Center, San Francisco, CA.

[Haworth co-indexing entry note]: "The Clinical Management of HIV Disease: Internet Resources on the World-Wide Web." Howe, J. Michael. Co-published simultaneously in *Health Care on the Internet* (The Haworth Press, Inc.) Vol. 2, No. 2/3, 1998, pp. 125-139; and: *HIV/AIDS Internet Information Sources and Resources* (ed: Jeffrey T. Huber) The Haworth Press, Inc., 1998, pp. 125-139; and: *HIV/AIDS Internet Information Sources and Resources* (ed: Jeffrey T. Huber) Harrington Park Press, an imprint of The Haworth Press, Inc., 1998, pp. 125-139. Single or multiple copies of this article are available for a fee from The Haworth Document Delivery Service [1-800-342-9678, 9:00 a.m. - 5:00 p.m. (EST). E-mail address: getinfo@haworthpressinc.com].

125

involved in the clinical treatment of HIV disease or the provision of information services would have been in a more difficult position. The biotechnology that made possible the laboratory diagnostics, the design of drugs, and the unraveling of the immune system provided significant insight into a complex and complicated biology and generated new knowledge at an incredible pace. In addition, the availability of this medical information through computer technologies facilitated access to an extensive body of literature.

This article describes selected Internet sites on the World-Wide Web that provide information about the clinical management of HIV disease. The amount of information on the Internet can be overwhelming, but the effective use of computer technology to locate authoritative up-to-date medical resources leads to a better knowledge of a very complex disease by both clinicians and patients. Although the debate continues about the reliability and credibility of information located on the Internet, the persistent user will soon develop skills to critically evaluate and analyze the material content. In addition, Internet sites and links change rapidly; therefore, some URLs may change before publication of this paper. All information, however, is current at the time this article was written (September 1997).

CLINICAL INFORMATION

Healthcare Communications Group Clinical Care Options for HIV (http://www.healthcg.com/)

Healthcare Communications Group (HCG) is a national organization dedicated to advancing medical knowledge for practical use by clinicians, patients, payers, and the broader health care industry. HCG provides CME and CEU accredited conferences, journals, monographs, and online continuing education programs. The Continuum of HIV Care Series for continuing medical education includes modules prepared by experts in their respective fields. Treatment issues and guidelines, conference summaries, HIV updates, and links are also available. Users can register for free e-mail announcements about new programs and updates.

Food and Drug Administration (FDA) (http://www.fda.gov)

The FDA home page is the beginning point to obtain information about the agency and drugs, food supplements, and medical devices the FDA regulates. One link (http://www.fda.gov/oashi/aids/hiv.htm) is specifically devoted to HIV and AIDS. Located here are sections on significant events in the epidemic (arranged chronologically), status of HIV/AIDS therapeutics, HIV testing, clinical trials and product development, evaluating medical therapies, barrier products, news releases and talk papers, articles and brochures, speeches, upcoming meetings, and other HIV Web sites.

HIVInSite (University of California, San Francisco) *(http://hivinsite.ucsf.edu/)*

The top-rated UCSF AIDS Program at San Francisco General Hospital (SFGH) and the UCSF Center for AIDS Prevention Studies created this site as a one-stop resource for reliable, peer-reviewed AIDS information. It is the only Web site in existence that contains research written, edited, and maintained by front-line AIDS researchers from a health sciences institution. Information is provided on treatment, clinical drug trials, epidemiology and basic research to social and policy issues, prevention programs, population subgroups, and ethics. Pages or subcategories include: prevention, medical, social issues, a U.S. map with state-by-state statistics, and key topics. HIVInSite's goal is to be a leading HIV Web site, not only because of its user-friendly design and the depth, scope, and quality of content, but also because it effectively ties together existing HIV resources on the Web. What is particularly unique is the availability of the AIDS Knowledge Base, a comprehensive text book on HIV disease from UCSF and SFGH for medical professionals and researchers.

HIV/AIDS Information (Doctor's Guide) *(http://www.pslgroup.com/aids.htm)*

This doctor's guide to the Internet is provided by P/S/L Consulting Group Inc., a Canadian organization dedicated to providing information services to help promote the informed and appropriate use of medicines by health care professionals and organizations as well as by the people to whom they are prescribed. To this end, Doctor's Guide was designed to help physicians have access to resources of the Internet and the World-Wide Web. A variety of links are available here that provide access to information about HIV disease.

HIV/AIDS Treatment Information Service (Public Health Service) *(http://www.hivatis.org/)*

The HIV/AIDS Treatment Information Service (ATIS) provides information about federally approved treatment guidelines for HIV/AIDS.

ATIS is staffed by bilingual (English and Spanish) health information specialists who answer questions on HIV treatment options using HSTAT, the National Library of Medicine database of HIV/AIDS treatment information, and other Federal resources. This site provides information on general treatment, drug treatment (e.g., protease inhibitors, approved drugs, viral load), HIV and women, and other resources including a glossary, related databases, and a chronology of HIV/AIDS. Links are available for other government and organization sites. The *AIDS Daily Summary* can also be accessed here. These abstracts from peer-reviewed journals and news services are provided as a public service by the CDC National Center for HIV, STD, and TB Prevention.

International Association of Physicians in AIDS Care (IAPAC) (http://www.iapac.org)

IAPAC provides a variety of information covering antiviral therapies, nutrition, pediatrics, opportunistic diseases, women's health issues, and information about and from IAPAC's journal. This page offers access to several conferences and provides full-text copies of information booklets and articles covering combination therapy, viral load management, protease inhibitors (available in seven languages), and a consumer's guide to prophylactic strategies for survival of HIV disease.

JAMA HIV/AIDS Information Center (http://www.ama-assn.org/special/hiv/hivhome.htm)

This site is an easy-to-use collection of high-quality resources for physicians, other health professionals, and the public. The site is made possible by an unrestricted educational grant from GlaxoWellcome and produced by *JAMA* staff under the direction of an editorial review panel and community advisory panel, but is intended as an informational resource only. Links allow access to clinical updates, news, and information on a broad range of social and policy issues.

The Johns Hopkins AIDS Service (http://www.hopkins-aids.edu/)

The goal of this site is to deliver a comprehensive, reliable, and timely source of information for HIV/AIDS care providers, with the hope that the information will be useful in the clinical management of HIV-infected patients. The site reflects Johns Hopkins AIDS Service dedication to both training care providers and keeping them as up-to-date as possible on new developments in the rapidly changing field of HIV/AIDS. Unique to this site is John Bartlett's comprehensive pocket guide to the care and management of patients with HIV and *The Hopkins HIV Report*, a bimonthly newsletter written by faculty members of the Schools of Medicine, Public Health, and Nursing who practice in the AIDS Service. Other sections explore and analyze the clinical and economic outcomes of treatment of HIV disease, epidemiological data, case rounds, women's issues, clinical trials, and events and conferences.

National Institute of Allergy and Infectious Diseases (Division of AIDS) (http://www.niaid.nih.gov/research/daids.htm)

The mission of the Division of AIDS is to increase knowledge of the pathogenesis, natural history, and transmission of HIV disease and to promote progress in its detection, treatment, and prevention. This is accomplished through planning, implementing, and evaluating programs in fundamental basic and clinical research, discovery and development of therapies for HIV infection and its complications, discovery and development of vaccines and other preventive interventions, and training of researchers in these activities. An extramural portfolio of grants and contracts addresses research in these areas. Located here are selections for upcoming meetings, conference/meeting summaries, research resources, DAIDS-supported programs (including the Adult AIDS Clinical Trials Group and the AIDS Clinical Trials Information Service) and publications.

CLINICAL TRIALS

AIDS Clinical Trials Information Service (U.S. Government) (http://www.actis.org/)

This service provides current information on federally and privately sponsored clinical trials for persons with AIDS and HIV infection. ACTIS English and Spanish-speaking health information specialists provide information on the purpose of each study, whether or not a study is open to enrollment, study locations, eligibility requirements and exclusion criteria, names and telephone numbers of contact persons, and drugs being studied. ACTIS maintains two online databases, one on trial protocols, the other drugs. The National Library of Medicine makes the information in these databases available to its worldwide network of users through two online databases, AIDSTRIALS and AIDSDRUGS. For information on federally approved treatment options, see the HIV/AIDS Treatment Information Service Web site described above. *The AIDS Daily Summary* can also be accessed here.

Canadian HIV Trials Network (CTN)
(http://www.hivnet.ubc.ca/ctn.html)

This is a federally funded, non-profit national organization created to facilitate HIV/AIDS clinical trial activity in Canada. The CTN is funded by Health Canada, and jointly sponsored by The University of British Columbia and St. Paul's Hospital, Vancouver. More information about CTN is located on this page as well as links to clinical trials, publications, and other resources.

CenterWatch Clinical Trials Listing Service
(http://www.centerwatch.com/)

You can use this listing to search for clinical trials, find out information about physicians and medical centers performing clinical research, and to learn about drug therapies newly approved by the Food and Drug Administration. You may also sign up for their mail notification service, which will inform you of future postings in a particular therapeutic area. Links include: clinical trials listing, patient notification service, profiles of centers conducting clinical research, newly approved drug therapies, industry providers, background information on clinical research, CenterWatch publications and services, and additional services.

HIV Clinical Trial Reviews (AIDS Action Committee of Massachusetts)
(http://library.jri.org/library/trials/aactrials/index.html)

These reviews are a part of the HIV Treatment Information Program, a reference service for people with HIV infection seeking up-to-date information about new treatments and drugs, side effects, clinical trials, and other HIV-related medical information. Through the AIDS Action Clinical Trials Reviews and the HIV Treatment Library (also available on this page), the goal of the program is to help people with HIV infection make informed decisions about their own treatment. The reviews are compiled by the HIV Resource Library.

Research and Clinical Trials (San Francisco General Hospital)
(http://sfghaids.ucsf.edu/ucsfresearch.html)

The University of California at San Francisco's AIDS Program at San Francisco General Hospital has been ranked number one for AIDS care among US hospitals. The laboratory and clinical research programs are dedicated to answering the most difficult scientific questions that the AIDS epidemic has posed and to increasing treatment options for people with HIV. The AIDS Program has the largest HIV/AIDS clinical research program in the Bay Area and this page provides up-to-date information about which studies are currently seeking volunteers. An online database can also be searched for HIV clinical trials in California. Also located here are answers to frequently asked questions about clinical trails.

CONFERENCES

Many of the HIV-related WWW pages have links to sites that provide information about national and international conferences. These are very useful, particularly when conferences are being held, as they offer immediate access to medical and scientific information related to the clinical management of HIV disease. Some of these sites include: IAPAC Conferences (http://www.iapac.org/conf.html); 4th Conference on Retroviruses and Opportunistic Infections (http://www.immunet.org/links/retroconf.html); 11th International Conference on AIDS (http://www.interchg.ubc.ca/aids11/aids96.html); and the 12th World AIDS Conference (http://www.aids98.ch/).

DATABASES-MEDICAL

National Library of Medicine (NLM): HIV/AIDS Resources (http://sis.nlm.nih.gov/aidswww.htm)

Online searching of NLM HIV-related databases is free and this site provides fact sheets that describe the databases and information that specifies the procedures for access to the databases. AIDSLINE is an online computer file containing references to the published literature on HIV infections and AIDS. It focuses on the biomedical, epidemiologic, health care administration, oncologic, and social and behavioral sciences' literature. The file contains citations (with abstracts if available) to journal articles, monographs, meeting abstracts and papers, government reports, theses, and newspaper articles from 1980 to the present. The AIDS-TRIALS (AIDS Clinical Trials) database provides information about AIDS-related studies of experimental treatments conducted under the Food and Drug Administration's investigational new drug regulations. AIDSTRIALS contains information about clinical trials of agents undergoing evaluation for the use of AIDS, HIV infection, and AIDS-related opportunistic diseases. After a search of AIDSTRIALS, a user may switch to the AIDSDRUGS database and view descriptive information about the agent used in the clinical trial.

National Library of Medicine: PubMed
(http://www.ncbi.nlm.nih.gov/PubMed/)

The announcement about free access to this experimental search system was made in June 1997. PubMed was developed in conjunction with publishers of biomedical literature as a search tool for accessing literature citations and linking to full-text journals at Web sites of participating publishers. The system provides access to the PubMed database of bibliographic information, which is drawn primarily from MEDLINE and preMEDLINE. In addition, for participating journals that are indexed selectively from MEDLINE, PubMed includes all articles from that journal, not just those that are included in MEDLINE.

PharmInfoNet Drug Database (DrugDB)
(http://pharminfo.com/drugdb/db_mnu.html)

DrugDB is a database of information about drugs that are listed alphabetically by generic names and tradenames in two separate sections. Each entry includes: generic name, tradename, manufacturer, therapeutic class, indication(s), and links to articles and archives on PharmInfoNet. A link is provided to an article on PharmInfoNet if the article mentions the drug and includes significant information about the drug. DrugDB is not an exhaustive listing of every drug on the market. It was developed by adding drugs that were reviewed during the approval process. Consequently, entries tend to be skewed toward new and investigational drugs. New drugs are continually added to the database.

DATABASES–NEWS SERVICES

AIDS Daily Summary (CDC National AIDS Clearinghouse)
(http://www.cdcnac.org/cgi/databases/news/adsdb.htm)

The AIDS Daily Summary Database consists of over 20,000 abstracts of articles about HIV/AIDS-related events in the news, trends in the epidemic, and research findings from major newspapers, wire services, medical journals and news magazines. New articles are abstracted and added to the database daily. The database includes articles from June 1988 to present. The user can search the database, view that day's summary, or view the last week of the summaries. These summaries, many of which are abstracts from the clinical literature, are particularly useful as they alert the health care practitioner to recent developments before a print copy of the journal is available.

HIV Newsline (AIDS Education Global Information System–AEGIS) (http://www.aegis.com/newslines.html)

AEGIS is arguably the most comprehensive source of information for HIV/AIDS on the Internet, and that applies to the news services that are available as well. Located here are databases for: AIDS Daily Summary, AIDS Weekly Plus, The Associated Press, Business Wire, The Chicago Tribune, The Los Angeles Times, The Miami Herald, PANOS, PR Newswire, Reuters, The San Francisco Chronicle, The San Francisco Examiner, The Wall Street Journal, and The Washington Post. For the most recent press reports regarding HIV disease, this should be the first site to search.

JAMA HIV/AIDS Newsline (http://www.ama-assn.org/special/hiv/newsline/newshome.htm)

This JAMA page offers daily updates, special reports, and background reports. Daily updates are available from Reuters Health Information Services and the Centers for Disease Control and Prevention AIDS News Summary. Special reports are in-depth articles from major professional sources. A collection of reports includes those on the science that underlies HIV/AIDS. Also included here are links that provide news coverage of recent conferences. What is particularly unique about this site is the analysis provided by JAMA's HIV/AIDS correspondent, Dennis Blakeslee, PhD. Dr. Blakeslee is, by training and experience, a journalist and science writer, a publications director, and a scientist. He has written extensively on a wide range of topics in biomedical research and health care, and is widely knowledgeable about HIV and AIDS.

JOURNALS/NEWSLETTERS

AIDS Online (http://www.aidsonline.com/)

AIDS Online is a peer-reviewed British publication that is published monthly, except semimonthly in March, July, and November, for a total of fifteen parts. *AIDS Online* offers full header and author abstracts in searchable text format of the journal on the day of acceptance by a peer-review panel, full text of current papers published electronically, and all published papers, fully searchable and maintained as a reference archive.

AIDS Information Newsletter
(gopher://gopher.niaid.nih.gov:70/11/aids/vaain)

The author of this article has published an electronic newsletter that has been transmitted to all medical centers within the VA system for almost seven years. The target audiences are health care practitioners, researchers, counselors/educators, librarians, and patients. Four series have been completed: HIV/AIDS in the Health Care Environment (16 parts); Tuberculosis and HIV Infection (19 parts); Women and HIV Infection (25 parts); Opportunistic Infections (32 parts); and Antiretroviral Therapy (ongoing series–20 parts completed). A Web page for the Center is currently under construction which will include this newsletter as well as additional information.

The AIDS Reader
(http://www.medscape.com/SCP/TAR/public/journal.TAR.html)

The *AIDS Reader*, published bimonthly by SCP Communications, Inc., is designed to provide clinicians with practical, scientifically sound information on the prevention, diagnosis, and treatment of HIV disease. By helping to bridge the gap between the specialist and the primary care physician, the goal is to assist clinicians caring for patients with this disease, and to help them improve the quality of life and treatment options for their patients. Scientific rigor is enforced through a process of peer review that evaluates information presented for fair balance, objectivity, independence, and relevance to educational need.

AIDS Treatment News Archive
(http://www.immunet.org/immunet/atn.nsf/homepage)

AIDS Treatment News is a primary treatment resource that provides information on public-policy developments, whether they involve changes in clinical trials or the drug-approval process, and news about the prevention of illnesses commonly associated with HIV. Published twice a month, the newsletter gives insights into treatments and options currently in use by physicians and other medical professionals as well as people with HIV disease. *AIDS Treatment News* is published twice a month by John S.

James, who is also the editor. All articles are fully searchable and indexed on this page. Also provided is John S. James' personal index.

Antiviral Agents Bulletin
(http://www.bioinfo.com/antiviral.html)

This publication is the only periodical specializing in antiviral drug and vaccine development and related activities. The *Bulletin* covers all antiviral therapeutics, but the largest portion of news articles, patents, etc., concerns HIV-infection and AIDS-related therapeutics. Articles describe and assess commercial and scientific developments, federal and regulatory activities, information resources, treatment advances and trends, patents and technology transfers.

Bulletin of Experimental Treatments for AIDS (BETA)
(http://www.sfaf.org/beta.html)

BETA, published quarterly by the Treatment Education and Advocacy Department of the San Francisco AIDS Foundation, covers new developments in AIDS treatment research. *BETA* publishes in-depth articles on treatment for HIV infection and AIDS-related illnesses for HIV positive individuals and their caregivers. Contributing writers include prominent researchers and clinicians as well as community advocates and activists. Each issue includes News Briefs, Research Notes, a Women and HIV/AIDS department, a listing of open clinical trials, and an extensive glossary of terms.

The Hopkins HIV Report
(http://www.hopkins-aids.edu/publications/index_pub.html)

This bimonthly newsletter is for practitioners who care for patients with HIV/AIDS. All articles are written by faculty of the Schools of Medicine, Public Health, and Nursing who practice in The Johns Hopkins AIDS Service. The most recent issue available on the page (July 1997) included articles on strategic antiretroviral therapy, adherence in the era of protease inhibitors, treatment of CMV retinitis, the Department of Health and Human Services draft guidelines on use of antiretroviral agents in HIV-infected adults.

Morbidity and Mortality Weekly Report (HIV/AIDS)
(http://www.cdc.gov/nchstp/hiv_aids/pubs/mmwr.htm)

This weekly publication by the Centers for Disease Control and Prevention is the primary source of information for CDC-approved guidelines, recommendations, and reports. This site in particular includes all reports in reverse chronological order related to HIV infection/AIDS for the past five years, beginning with the current year. This is particularly useful for clinicians who are searching for CDC published information for the clinical management of HIV disease.

Other Publications

Other useful newsletters include: *GMHC Treatment Issues* (http://www. gmhc.org/aidslib/ti/ti.html) and *Project Inform Perspectives* (http://www. projinf.org/pub/pip_index.html). Access to several peer-reviewed journals is available on the American Medical Association Publishing Home Page (http://www.ama-assn.org/scipub.htm). Additional journal titles and locations are: *The Lancet* (http://www.thelancet.com/); *Nature Medicine* (http:// medicine.nature.com/); *New England Journal of Medicine* (http://www. nejm.org/); and *Science* (http://www.sciencemag.org/).

REFERENCE RESOURCES

AIDS Knowledge Base
(http://hivinsite.ucsf.edu/akb/1997/)

This comprehensive textbook on HIV disease from the University of California, San Francisco, and the San Francisco General Hospital for medical professionals and researchers includes chapters on: transmission, testing, natural history, clinical spectrum and general management of HIV disease, and clinical management. The complete 1994 edition plus new chapters from the unpublished 1997 edition are available.

Medical Management of HIV Infection
(http://www.hopkins-aids.edu/publications/index_pub.html)

This text is by Johns Hopkins AIDS Service and has now been accepted as the standard of care for quality of assurance by Maryland Medicaid. Originally written in 1989 as a 28-page pamphlet, the first edition summarized most of the necessary information available about HIV infection at that time. The 1997 edition incorporates the recent changes in the field and required a nearly complete revision of the 1996 edition. Perhaps the greatest asset of this book is the effort and methods used to keep it topical. This book is revised and updated annually in a print version; new developments that dictate changes in management strategies are reviewed throughout the year in *The Hopkins HIV Report* (described earlier under the heading, Journals/Newsletters). Now that the book is published electronically, the online version will be updated monthly, so it will be possible to immediately incorporate findings presented at a recent conference or groundbreaking news announced recently. This process of continual update ensures that this text will provide clinicians with the most relevant and timely information on the clinical care of patients with HIV infection.

Merck Manual (1992)
(http://www.merck.com/!!rIkrg3k5NrIkrg3k5N/pubs/mmanual/)

This manual is expressly designed to meet the needs of general practitioners in selecting medications and has become a highly valued resource for medical students and clinical staff. It is the most widely used medical text in the world and its primary purpose remains the same–to provide useful clinical information to practicing physicians, medical students, interns, residents, and other health care professionals. The first section of the sixteenth edition (1992) covers problems associated with infectious diseases. The link to human immunodeficiency virus infection includes sections on etiology and pathogenesis, epidemiology, symptoms and signs, laboratory findings and diagnosis, prognosis, prevention, and treatment.

Glossaries

Several useful glossaries are located on the Internet including those provided by the AIDS Treatment Data Network (http://www.aidsnyc.org/network/drugloss.html); Gay Men's Health Crisis (http://www.critpath.org/research/gmhgloss.htm); and the American Medical Association Home Page (http://www.ama-assn.org/special/hiv/glossary/gloshome.htm).

THERAPY–ANTIRETROVIRAL

The Food and Drug Administration has approved several drugs for antiretroviral therapy. These include:

1. Nucleoside Analogs–Reverse Transcriptase Inhibitors: AZT (zidovudine, azidothymidine, Retrovir); ddI (didanosine, Videx–approved October 1991); ddC (deoxycytidine, zalcitabine, Hivid); d4T (stavudine, Zerit); 3TC (lamivudine, Epivir).
2. Non-Nucleoside Analogs–Reverse Transcriptase Inhibitors: nevirapine (Viramune); delavirdine (Rescriptor).

3. Protease Inhibitors: saquinavir (Invirase); ritonavir (Norvir); indinavir (Crixivan); nelfinavir (Viracept).

One of the most useful sites to obtain comprehensive, timely information about antiretroviral therapies is the Antiviral Therapies section (http://www.iapac.org/protidx.html) on the International Association of Physicians in AIDS Care page. At the time of this writing, this link provided articles from the IAPAC journal covering antiviral therapies, recommendations for monitoring viral load in clinical practice, drug levels and the generation of resistance, summary reports for three conferences, and consumer and patient information. The latter includes a consumer's guide to prophylaxis strategies for survival of HIV disease and two booklets published by IAPAC, one on combination therapy and another that describes protease inhibitors (available in German, English, Portuguese, Spanish, Italian, French and Japanese).

Several pharmaceutical companies offer information about their antivirals: Agouron (Viracept) (http://www.agouron.com/htdocs/viracept2.html); Glaxo Wellcome (AZT) (http://www.glaxowellcome.co.uk/home.html); Merck (Crixivan) (http://www.merck.com); and Roxane (Viramune) (http:// www.viramune.com/). Stadtlanders Pharmacy (http://www.stadtlander.com/hivfocus.htm) also has an informative page. Stadtlanders offers medication delivery and compliance, education materials, and insurance billing. The pharmacy has underwritten the cost of Raise the Rainbow project. It also provides HIV nutrition seminars around the country, LIFETIMES magazine, other education and wellness materials, a national AIDS resource directory, and HIV Nutrition Guidelines.

VIRTUAL LIBRARIES

All of the Web links described thus far have been included because they provide information specifically related to the clinical management of HIV disease. Several Pages, however, have links to other resources that provide content not only for the medical management of HIV disease, but also for virtually all aspects of HIV/AIDS. These are invaluable lists for clinicians who want a comprehensive coverage of the disease. These include the following:

- AIDS and HIV (Galaxy)
  (http://galaxy.einet.net/galaxy/Community/Health/Diseases/AIDS-and-HIV.html)
- AIDS Education Global Information System (AEGIS)
  (http://www.aegis.com/)

- AIDS Net-Information and Connections
  (http://www.treknet.net/html/research/aids.shtml)
- AIDS Resource List (Celine's)
  (http://www.teleport.com/~celinec/aids.shtml)
- AIDS Treatment News Internet Directory
  (http://www.aidsnews.org/)
- AIDS Treatment News Internet Directory 2
  (http://www.aidsnews.org/atnid2.html)
- AIDS Treatment Publications (Critical Path AIDS Project)
  (http://www.critpath.org/pubs.htm)
- AIDS Virtual Library (Planet Q)
  (http://planetq.com/aidsvl/index.html)
- Health InfoWeb (JRI)
  (http://www.infoweb.org/)
- HIV/AIDS Outreach Project (Vanderbuilt University)
  (http://www.mc.vanderbilt.edu/adl/aids_project/)
- HIV/AIDS Resource Sampler (NN/LM PNR)
  (http://www.nnlm.nlm.nih.gov/pnr/etc/aidspath.html)
- MedWeb: AIDS and HIV
  (http://www.gen.emory.edu/medweb/medweb.aids.html)
- New York Online Access to Health (NOAH)
  (http://www.noah.cuny.edu/aids/aids.html)
- Yahoo!-AIDS/HIV
  (http://www.yahoo.com/Health/Diseases_and_Conditions/
  AIDS_HIV)

Antiretroviral Drug Development and Information on the Internet

Ed Casabar

SUMMARY. The rapid growth of the Internet parallels advances in the treatment of primary HIV infection. Recent breakthroughs in the use of viral load testing have shifted the focus away from monotherapy to highly active antiretroviral therapy (HAART) with combination regimens. Currently, three classes of drugs are available for the treatment of primary HIV infection and include the nucleoside and non-nucleoside reverse transcriptase inhibitors and the protease inhibitors. A review of the historical development of these compounds and some Internet resources is presented. *[Article copies available for a fee from The Haworth Document Delivery Service: 1-800-342-9678. E-mail address: getinfo@haworthpressinc.com]*

In 1987, when zidovudine was approved by the Food and Drug Administration (FDA) for the treatment of human immunodeficiency virus (HIV) infection, the World-Wide Web had not yet been created. The rapid growth of the Web, however, parallels the speed at which new antiviral drugs are being investigated. To stay abreast with this rapidly changing field, many clinicians and patients are searching for information on the Internet. This article reviews the historical development and some complexities of treating primary HIV infection and provides some resources of information available on the Internet for both patients and their caregivers.

Ed Casabar, PharmD, BCPS (casabar@osler.wustl.edu), is Infectious Diseases Clinical Pharmacist, Department of Pharmacy, Barnes-Jewish Hospital, St. Louis, MO.

[Haworth co-indexing entry note]: "Antiretroviral Drug Development and Information on the Internet." Casabar, Ed. Co-published simultaneously in *Health Care on the Internet* (The Haworth Press, Inc.) Vol. 2, No. 2/3, 1998, pp. 141-151; and: *HIV/AIDS Internet Information Sources and Resources* (ed: Jeffrey T. Huber) The Haworth Press, Inc., 1998, pp. 141-151; and: *HIV/AIDS Internet Information Sources and Resources* (ed: Jeffrey T. Huber) Harrington Park Press, an imprint of The Haworth Press, Inc., 1998, pp. 141-151. Single or multiple copies of this article are available for a fee from The Haworth Document Delivery Service [1-800-342-9678, 9:00 a.m. - 5:00 p.m. (EST). E-mail address: getinfo@haworthpressinc.com].

DRUG APPROVAL PROCESS

The premarket approval of new drugs is dependent upon the scientific evaluation of safety and efficacy by the FDA. Clinical trials may be independently sponsored by pharmaceutical companies or conducted as part of federally funded projects. In the United States, the National Institutes of Health (NIH) coordinates the federally funded scientific efforts of the AIDS Clinical Trials Group (ACTG). The ACTG research centers (http://www.niaid.nih.gov/research/Daids.htm) are dedicated to critically evaluating new AIDS therapies through the implementation of large, well-designed drug studies. Patients and caregivers can find additional information on federally and privately funded clinical trials of new AIDS drugs at the AIDS Clinical Trials Information Service (http://www.actis.org/).

Bringing a drug to market, however, has traditionally been a long and arduous process. Before 1987, it would take, on average, eight and one-half years to study and test a new drug before the FDA could approve it for the general public.[1] During the AIDS epidemic, activists have criticized the FDA for its lengthy review process. One critique of the FDA's responsiveness to the AIDS epidemic is available at the Gay Men's Health Crisis (http://www.gmhc.org/aidslib/ti/ti1001/ti1001a.html).

In acknowledgment of its critics, the FDA has made advances in expediting drug applications. On May 22, 1987, the FDA took its first steps by creating the "accelerated approval" priority category for new drugs. This new regulation gave promising AIDS drugs the highest priority for FDA review. On December 11, 1992, the FDA passed its final regulation on accelerated approval of drugs. This revision established that surrogate endpoints of efficacy could be used as a basis for FDA approval. An explanation of the expedited approval process can be found at the FDA (http://www.fda.gov/oashi/aids/expanded.html). A consumer-oriented publication is also available from the FDA (http://www.fda.gov/fdac/special/newdrug/ndd_toc.html). Recent legislation affecting the drug approval process can be searched at Thomas, the legislative information Internet site of the United States Congress (http://thomas.loc.gov/).

VIRAL LIFE CYCLE

By determining key sequences in the life cycle of HIV, research into antiviral drugs was made possible. HIV is a retrovirus, unique among viruses because its replication requires the incorporation of its genetic material (RNA) into the DNA of the human it infects. To accomplish this, HIV possesses three unique enzymes: reverse transcriptase, integrase, and protease. For now, FDA-approved drug therapy for primary HIV infection is targeted at inhibiting reverse transcriptase and protease. A description of HIV replication and the sites for drug activity can be found at Cornell University (http:// edcenter.med.cornell.edu/CUMC_PathNotes/HIV_Infection/ HIV_Infection_TOC.html).

NUCLEOSIDE ANALOGS

In 1985, zidovudine (ZDV, AZT, Retrovir) was the first compound found to have antiretroviral activity. A member of the nucleoside reverse transcriptase inhibitor (nRTI) family, zidovudine was approved by the FDA in March 1987. Though ZDV represented the first hope that HIV could be controlled, ZDV was not without toxicity. As many as 30% of patients develop dose-related anemia or neutropenia. Another serious, though less common adverse effect is myopathy.[2] The search for less toxic therapies led to the development of other nucleoside analogs. It was not until 1991, however, that a second nRTI, didanosine (DDI, Videx), became commercially available. A related compound, zalcitabine (DDC, Hivid), would be approved in 1992. Though DDI and DDC were promoted initially as alternatives for those patients not tolerating ZDV, their toxicities may also be serious and include painful peripheral neuropathies and pancreatitis. The most recently marketed nRTIs include stavudine (D4T, Zerit), which received FDA approval in 1993, and lamivudine (3TC, Epivir), which was approved in December 1995. Table 1 lists the medications that are currently approved by the FDA for treating HIV infection.

Despite the early advances seen with ZDV and other nRTIs, the search for more effective drugs continued. The worldwide scientific community comes together biennially at the International AIDS Conferences, in part, to showcase exciting breakthroughs in drug therapies. Unfortunately, when the 1994 International AIDS Conference was held in Yokahama, Japan, the mood of many attendees was reserved. At that time, data from large, well-designed clinical trials suggested that the benefit of nRTI monotherapy was not durable, and lasted only 12 to 24 months in asymptomatic patients.[3-5] An archive of the abstracts from the Yokohama Conference can be found via a Gopher search (http://gopher.hivnet.org:70/1/ magazines/abst).

Questions regarding the durability of nRTI monotherapy spurred research into the use of combination regimens containing two nucleoside analogs. Two large, randomized clinical trials of dual nRTI combination therapy, ACTG 175 and the European Delta Study, confirmed that nRTI monotherapy was inferior to combination therapy in terms of disease

TABLE 1. Antiretroviral Therapy Approved by the FDA*

Class	Generic Name	Other Names	Manufacturer	FDA Approval
Neucleoside reverse transcriptase inhibitors (nRTI)	Zidovudine	ZDV, AZT, Retrovir®	Glaxo-Wellcome	March 1987
	Didanosine	DDI, Videx®	Bristol Meyers-Squibb	October 1991
	Zalcitabine	DDC, Hivid®	Hoffman-La Roche	June 1992
	Stavudine	D4T, Zerit®	Bristol Meyers-Squibb	June 1994
	Lamivudine	3TC, Epivir®	Glaxo-Wellcome	November 1995
Non-nucleoside reverse inhibitors (nnRTI)	Nevirapine	Viramune®	Boehringer Ingelheim/ Roxane	June 1996
	Delavirdine	Rescriptor®	Pharmacia/ UpJohn	April 1997
Protease inhibitors	Saquinavir	Invirase®	Hoffman-LaRoche	December 1995
	Ritonavir	Norvir®	Abbott	March 1996
	Indinavir	Crixivan®	Merck	March 1996
	Nelfinavir	Viracept®	Agouron	March 1997

*Current as of September 1997

progression and survival.[6-7] These studies profoundly affected the philosophy of antiretroviral therapy of the time and eventually, nRTI monotherapy was no longer considered an acceptable regimen.

NON-NUCLEOSIDE ANALOGS

A second class of compounds is the non-nucleoside reverse transcriptase inhibitors (nnRTIs). Nevirapine (Viramune) was the first of these drugs and was approved in June 1996. Though generally well-tolerated, nevirapine commonly produces skin rash. Non-nucleoside analogs must also be used in combination with other drugs to prevent the rapid development of resistance with nnRTI monotherapy. Information about nevirapine can be found at the manufacturer's Web site (http://www.viramune.com/). A second nnRTI, delavirdine (Rescriptor), received FDA approval in April 1997 amid political controversy. AIDS activists have held that patients should be given as many therapeutic options as possible. Others have questioned the utility of delavirdine since HIV quickly develops resistance. A discussion of the controversial FDA approval of delavirdine can be found at The Body Health Resources Corporation (http://www.thebody.com/gmhc/issues/aprmay97.html#eleven).

VIRAL DYNAMICS

The recent advances in antiretroviral therapy were partly a result of a paradigm shift in the understanding of the dynamics of HIV infection. Until 1993, it was believed that after primary infection, HIV lay dormantly in the human host for many years until some triggering event would initiate rapid viral production and the eventual destruction of the immune system.[8] With the evolution of highly sensitive viral assays based on the polymerase chain reaction (PCR), researchers measured the number of viral particles in plasma (viral load). Using these very sensitive assays, Ho and Wei showed that HIV infection was in fact prolific and continuous and not quiescent as previously believed.[9-10] A summary of viral load assays can be found at the University of San Francisco HIV InSite (http://hivinsite.ucsf.edu/akb/1994/2-4/index.html#2-4-A).

Uncovering a relationship between viral load and HIV disease progression was the next important milestone. Mellors et al. published an analysis of plasma samples obtained from HIV patients enrolled in the Multicenter AIDS Cohort Study (MACS). The MACS data suggested that an individual's baseline viral load, or set point, was a better predictor of long-term survival than other laboratory tests or surrogate markers (i.e., CD4 T-lymphocyte count).[11] Theoretically, modifying a patient's set point could alter a patient's prognosis.

PROTEASE INHIBITORS

The important implications of viral load and its effect on long-term prognosis focused more attention to finding new drugs that could profoundly inhibit viral replication. One candidate class of drugs, the protease inhibitors, was developed after the chemical structure of protease was elucidated by x-ray crystallography.[12] Protease is a viral enzyme responsible for cleaving viral polypeptides produced during translation of viral genes. Though the protease inhibitors have marked effects on viral replication, their antiviral effects are short-lived, since HIV resistance to monotherapy quickly develops.[13] To overcome drug resistance, drug combinations or "drug cocktails" have been devised. Originally, these combinations consisted of a protease inhibitor and two nRTIs. Combinations of nnRTIs and nRTIs, with or without protease inhibitors, are being investigated.

In July 1996, when the Eleventh International AIDS Conference convened in Vancouver, British Columbia, experience with viral load testing and double and triple drug regimens had grown exponentially. These novel drug combinations acquired a new acronym–HAART (highly active antiretroviral therapy)–and appeared powerful enough to lower viral burden below the limits of detection. In contrast to the mood at the Yokohama Conference in 1994, scientists at the 1996 Vancouver Conference were more hopeful than ever that HIV infection could not only be controlled, but also possibly be eradicated from an infected individual's body. Abstracts and editorial discussions of the Vancouver Conference can be found at several Web sites including: Healthcare Communications Group (http:// 207.226.163.175/hiv/conference/conference.html), the National Library of Medicine (http://sis.nlm.nih.gov/aidsabs.htm), and the American Medical Association (http://www.ama-assn.org/special/hiv/xi-conf/aids11.htm).

Not surprisingly, data presented at the Vancouver Conference have convinced many caregivers and patients to accept HAART as a new standard of care. Saquinavir (Invirase) was FDA approved in December 1995, followed by ritonavir (Norvir) and indinavir (Crixivan) in March 1996. Nelfinavir (Viracept) received FDA approval in March 1997.

PRACTICAL APPLICATION

With eleven antiretroviral drugs now available commercially, the application of HAART in routine clinical practice has become extremely complex. For many patients, protease inhibitors produce a wide range of toxicities. The protease inhibitors and non-nucleoside analogs may also interact with many drugs that share metabolic pathways dependent upon cytochrome P450 hepatic enzymes. The result is an alteration in plasma drug concentrations. In addition, some protease inhibitors require proper timing of administration with meals, adequate fluid hydration, and even attention to the temperature and humidity of drug storage. Patients must also face the costly reality of paying for these complex drug regimens and their monitoring.

To illustrate these difficulties, a patient on a "cocktail" of ZDV, 3TC, and indinavir ingests a total of 10 capsules a day. ZDV and 3TC can be taken without regard to food. However, since its absorption is affected by meals, indinavir should be taken three times daily on an empty stomach (one hour before a meal or two hours after). For many patients, the careful timing separation of meals and indinavir is difficult. Patients taking indinavir must also drink approximately 1.5 liters of fluid daily to prevent the development of indinavir-induced kidney stones. A patient's average estimated yearly drug expenditure for this triple drug regimen approaches $12,000.[14] Patients who need financial assistance with medications may find information at The Access Project (http://204.179.124.69/network/access/index.html).

For those with advanced HIV disease complicated by concomitant opportunistic infections, the addition of antibiotics to an existing HAART regimen only adds additional problems. Antibiotics may interact with antiretroviral drugs to produce additive toxicities (e.g., ganciclovir and ZDV-induced anemia) or lower the antiretroviral drug efficacy via the induction of hepatic enzymes. One important drug interaction occurs with rifampin. This anti-tuberculous drug is a potent inducer of the Cyp3A4 subclass of cytochrome P450 hepatic oxidases and results in subtherapeutic protease inhibitor plasma concentrations. Recently, the Centers for Disease Control and Prevention (CDC) published their recommendations on managing the drug interactions associated with HAART and anti-tuberculous therapy.[15] The CDC recommendations are also available on the Internet via anonymous FTP (ftp://ftp.cdc.gov/pub/Publications/mmwr/wk/mm4542.pdf). A broader discussion of adverse drug effects and interactions is also available at the Critical Path (http://www.critpath.org/newsletters/critpath/31drgint.htm).

The novelty and rapid approval of antiretroviral drugs present another potential problem. Adverse effects that were not found in early clinical trials are now becoming evident as combination regimens become widely used. As an example, in June 1997, the FDA released a warning that the protease inhibitors produced hyperglycemia or diabetes mellitus in 84 patients (http://www.fda.gov/bbs/topics/ANSWERS/ANS00800.html).

The complexities of HAART can be overwhelming for many patients. Experts agree that HAART can only be successfully applied in the face of proper patient education and motivation. Patient education materials can be found on the World-Wide Web. Useful resources include those from Project Inform (http://www.projinf.org/), the American Medical Association (http://www.ama-assn.org/insight/spec_con/hiv_aids/crix.htm), Medconsult. com (http://www.mediconsult.com/frames/aids/support/), the Pharmaceutical Information Network (http://www.pharminfo.com/pharmin.html) and the University of San Francisco HIV InSite (http://arvdb.ucsf.edu/ all-drug.cfm). A pictorial guide to HIV drugs that includes color pictures of dosage forms can be found at the Department of Pharmacy, NIH Clinical Center (http://www.cc.nih.gov/phar/hiv_mgt/index.html).

TREATMENT GUIDELINES

Faced with the challenges of properly prescribing and monitoring complex antiretroviral regimens, several organizations have established treatment guidelines that summarize the proper implementation of HAART. The International AIDS Society (IDSA)-USA was the first group to publish a set of recommendations.[16] Their consensus statement is also available at the JAMA Web pages (http://www.ama-assn.org/special/hiv/ library/jama97/st7009.htm). Additional guidelines from the NIH and the Department of Health and Human Services (DHHS) were published on the Internet in draft form for public review. The NIH/DHHS drafts can be found at the HIV/AIDS Treatment Information Service (http://www. hivatis.org/guidelin.html). Both the IDSA and NIH/DHHS guidelines outline the importance of frequent viral load monitoring to determine the effectiveness of HAART. Expert interpretations of the IDSA and NIH/ DHHS guidelines can be found at the International Association of Physicians in AIDS Care Antiviral Page (http://www.iapac.org/protidx.html), Johns Hopkins University (http://www.hopkins-aids.edu/), the Treatment Action Group, New York (http://www.aidsnyc.org/tag/phs.html), and Washington University, St. Louis (http://id.wustl.edu/~actu/).

The IDSA and the United States Public Health Service (USPHS) have also recently published guidelines for the prevention of opportunistic infections associated with HIV infection.[17] These may be also downloaded from the CDC via anonymous FTP (ftp://ftp.cdc.gov/pub/Publications/ mmwr/rr/rr4612.pdf).

EVOLVING SCIENCE

Although the IDSA and NIH/DHHS consensus statements represent the current state of the art in antiretroviral therapy, this area of medicine is rapidly evolving. Many developments continue to be discussed at national meetings including the January 1997 Conference on Retroviruses and Opportunistic Infections in Washington, D.C. A summary of the Washington, D.C. Conference can be found at Healthcare Communications Group (http://207.226.163.175/hiv/4thconf/).

A few promising investigational drugs include the nRTI, 1592-U89, the nnRTI, DMP-266, and a soft gelatin capsule formulation of saquinavir that appears to have improved bioavailability. The exact role of these investigational drugs has yet to be determined. Information regarding investigational drugs can be found at the University of San Francisco HIV InSite (http://hivinsite.ucsf.edu/tsearch) and the National AIDS Treatment Advocacy Project (http://www.aidsinfonyc.org/natap/conf/index.html).

Drug resistance is an important emerging field of study. Resistance may occur from the natural and spontaneous mutations of protease or reverse transcriptase. Mutations may also develop when antiretroviral therapy is inadequately managed or not potent enough to suppress viral replication adequately. Without careful adherence to proper dosing, timing of administration, and attention to drug interactions, plasma protease inhibitor concentrations may be lowered. This only increases the likelihood those drug resistance mutations will occur. Scientists are now refining assays that can determine the genotypic and phenotypic patterns of HIV resistance. In addition, drug sensitivity assays are being investigated.[18-19] A review of drug resistance can be found at the American Medical Association Archives (http://www.ama-assn.org/special/hiv/library/vella.htm). Recent advances in this field were presented at the June 1997 International Workshop on Drug Resistance, Treatment Strategies and Eradication in St. Petersburg, Florida. Abstracts from the Workshop are on the Web (http://207.226. 163.175/hiv/hivresistance/).

Patients and caregivers who want to stay abreast with the rapid changes in antiretroviral therapy can do so via the Internet. Breaking developments can be found at several news services including the AIDS Education Information Service (http://www.aegis.org/), the Biotechnology Information Institute Antiviral Agents Bulletin (http://www.bioinfo.com/biotech/antiviral.html), and Usenet (news:sci.med.aids). A searchable database of articles from the AIDS Treatment Network is available at Immunet (http://www.immunet.org/).

RELIABILITY OF INFORMATION

Although the Internet has become an easily accessible source of information on drug therapy for patients and their caregivers, concerns have been raised. The FDA, pharmaceutical industry, and organizations representing caregivers and patients have taken interest in improving the reliability of Internet-based drug information and preventing fraudulent claims from being disseminated. A discussion of this topic is available at the Internet Health Care Coalition (http://pharminfo.com/conference/ihc_

hp.html) and its affiliated Listserv discussion group, MedWebMasters-L (http://pharminfo.com/conference/MWM-L.html).

Access to information via the Internet can be a very useful tool for staying abreast with emerging topics in drug therapy. However, consumers should be aware that some information on the Internet may not always be credible since the medium is currently uncensored and uncontrolled.

REFERENCES

1. *From Test Tube To Patient, New Drug Development In The United States.* 2nd ed. Rockville, MD: US Food and Drug Administration, 1995.

2. Rachlis A. et al. "Zidovudine Toxicity: Clinical Features And Management." *Drug Safety* 8 (April 1993): 312-20.

3. Volberding P.A. et al. "Zidovudine In Asymptomatic Human Immunodeficiency Virus Infection: A Controlled Trial In Persons With Fewer Than 500 CD4-Positive Cells Per Cubic Millimeter." *N Engl J Med* 322 (April 5, 1990): 941-9.

4. Volberding P.A. et al. "The Duration Of Zidovudine Benefit In Persons With Asymptomatic HIV Infection: Prolonged Evaluation Of Protocol 019 Of The AIDS Clinical Trials Group." *JAMA* 272 (August 10, 1994): 437-442.

5. Concorde Coordinating Committee. "Concorde: MRC/ANRS Randomised Double-Blind Controlled Trial Of Immediate And Deferred Zidovudine In Symptom-Free HIV Infection." *Lancet* 343 (April 9, 1994): 878-81.

6. Hammer S.M. et al. "A Trial Comparing Nucleoside Monotherapy With Combination Therapy In HIV-Infected Adults With CD4 Cell Counts From 200 To 500 Per Cubic Millimeter." *N Engl J Med* 335 (October 10, 1996): 1081-90.

7. Delta Coordinating Committee. "Delta: A Randomised Double-Blind Controlled Trial Comparing Combinations Of Zidovudine Plus Didanosine Or Zalcitabine With Zidovudine Alone In HIV-Infected Individuals." *Lancet* 348 (August 3, 1996): 283-91.

8. Pantaleo G. et al. "New Concepts In The Immunopathogenesis Of Human Immunodeficiency Virus Infection." *N Engl J Med* 328 (February 4, 1993): 327-35.

9. Ho D.D. et al. "Rapid Turnover Of Plasma Virions And CD4 Lymphocytes In HIV-1 Infection." *Nature* 373 (January 12, 1995): 123-6.

10. Wei X. et al. "Viral Dynamics In Human Immunodeficiency Virus Type-1 Infection." *Nature* 373 (January 12, 1995): 117-22.

11. Mellors J.W. et al. "Prognosis In HIV-1 Infection Predicted By Quantity Of Virus In Plasma." *Science* 272 (May 24, 1996): 1167-70.

12. Jhoti H. et al. "X-Ray Crystallographic Studies Of A Series Of Penicillin-Derived Asymmetric Inhibitors Of HIV-1 Protease." *Biochemistry* 33 (July 19, 1994): 8417-27.

13. Molla A. et al. "Ordered Accumulation Of Mutations In HIV Protease Confers Resistance To Ritonavir." *Nature Med* 2 (July 1996): 760-6.

14. Cardinale V. et al. *Drug Topics Red Book*. New Jersey: Medical Economics Company, 1997.

15. "Clinical Update: Impact Of HIV Protease Inhibitors On The Treatment Of HIV-Infected Tuberculosis Patients With Rifampin." *MMWR* 45 (October 25, 1996): 921-5.

16. Carpenter, C.C. et al. "Antiretroviral Therapy For HIV Infection In 1997: Recommendations Of The International AIDS Society-USA Panel." *JAMA* 277 (June 25, 1997): 1962-9.

17. "1997 USPHS/IDSA Guidelines For The Prevention Of Opportunistic Infections In Persons Infected With Human Immunodeficiency Virus." *MMWR* 46 (June 27, 1997): 1-46.

18. Mayers D. "Rational Approaches To Resistance: Nucleoside Analogues." *AIDS* 10 (Supplement 1, November 1996): S9-13.

19. Condra J.H. et al. "In Vivo Emergence Of HIV-1 Variants Resistant To Multiple Protease Inhibitors." *Nature* 374 (April 6, 1995): 569-71.

Index

ABIA (Associacao Brasileira
  Interdisciplinar de AIDS),
  71
Access Project, 147
Acquired immunodeficiency
  syndrome. *See* AIDS
ACTG (AIDS Clinical Trials Group),
  66,95,142
  Adult, 129
ACTIS (AIDS Clinical Trials
  Information Service),
  93,100,102-103,104,128,
  129-130,142
Active Digital Library, HIV/AIDS
  Project, 75
ACT UP, 117
Acupuncture.com, 121
Adolescents, HIV/AIDS Internet
  resources for, 42-43,53-55,
  57-60,72
Adult AIDS Clinical Trials Group,
  129
AEGIS: AIDS Education Global
  Information System, 50,83,
  138
AETC (AIDS Education and
  Training Center for Texas
  and Oklahoma), 27,28,29,
  32,35,36
African Americans
  AIDS epidemic among, 40,42,
  43-44
  HIV/AIDS Internet information
  for, 43-44
  WAM Foundation, Inc. for, 29,
  31,38
African-American women,
  HIV/AIDS in, 63,66,67
  adolescents, 42

deaths related to, 49
Agency for Health Care Policy and
  Research, 93-94,103
AID Atlanta, 12-13,14,19-20
AIDS (acquired immunodeficiency
  syndrome), first recognition
  of, 12
AIDS Action Committee of
  Massachusetts, 12-13,18-19
AIDS Alternative Treatment, 114
AIDS and HIV (Galaxy), 138
AIDS Care Clinical Research
  Program, 33
AIDS cases, number of
  among African Americans, 42,
  43-44
  in Houston, Texas, 26
  in Louisville, Kentucky, 7
  in rural areas, 47
  underreporting of, 79
  in United States, 79
  worldwide, 21-22
*AIDS Clinical Trial Results
  Database*, 103
AIDS Clinical Trials Group (ACTG),
  66,95,142
  Adult, 129
AIDS Clinical Trials Information
  Service (ACTIS), 93,100,
  102-103,104,128,129-130,
  142
AIDS Coalition of Coastal Texas,
  Inc., 33,35-36
AIDSDACT, 84
AIDS Daily Summary,
  31,75,83,101,128,130,
  132-133
AIDS-defining conditions, 65
AIDSDRUGS,

153

Haworth
DOCUMENT DELIVERY
SERVICE

This valuable service provides a single-article order form for any article from a Haworth journal.

- *Time Saving:* No running around from library to library to find a specific article.
- *Cost Effective:* All costs are kept down to a minimum.
- *Fast Delivery:* Choose from several options, including same-day FAX.
- *No Copyright Hassles:* You will be supplied by the original publisher.
- *Easy Payment:* Choose from several easy payment methods.

Open Accounts Welcome for . . .
- Library Interlibrary Loan Departments
- Library Network/Consortia Wishing to Provide Single-Article Services
- Indexing/Abstracting Services with Single Article Provision Services
- Document Provision Brokers and Freelance Information Service Providers

---

MAIL or *FAX* THIS ENTIRE ORDER FORM TO:

Haworth Document Delivery Service
The Haworth Press, Inc.
10 Alice Street
Binghamton, NY 13904-1580

or FAX: 1-800-895-0582
or CALL: 1-800-429-6784
9am-5pm EST

PLEASE SEND ME PHOTOCOPIES OF THE FOLLOWING SINGLE ARTICLES:

1) Journal Title: _____

Vol/Issue/Year: _____ Starting & Ending Pages: _____

Article Title: _____

_____

2) Journal Title: _____

Vol/Issue/Year: _____ Starting & Ending Pages: _____

Article Title: _____

_____

3) Journal Title: _____

Vol/Issue/Year: _____ Starting & Ending Pages: _____

Article Title: _____

_____

4) Journal Title: _____

Vol/Issue/Year: _____ Starting & Ending Pages: _____

Article Title: _____

_____

(See other side for Costs and Payment Information)

COSTS: Please figure your cost to order quality copies of an article.

1. Set-up charge per article: $8.00
    ($8.00 × number of separate articles)   _____

2. Photocopying charge for each article:

    1-10 pages: $1.00   _____

    11-19 pages: $3.00   _____

    20-29 pages: $5.00   _____

    30+ pages: $2.00/10 pages   _____

3. Flexicover (optional): $2.00/article   _____

4. Postage & Handling: US: $1.00 for the first article/
    $.50 each additional article   _____

    Federal Express: $25.00   _____

    Outside US: $2.00 for first article/
    $.50 each additional article _____

5. Same-day FAX service: $.50 per page   _____

                    GRAND TOTAL: _____

METHOD OF PAYMENT: (please check one)

❏ Check enclosed   ❏ Please ship and bill. PO # _____
                    (sorry we can ship and bill to bookstores only! All others must pre-pay)

❏ Charge to my credit card: ❏ Visa; ❏ MasterCard; ❏ Discover;
                    ❏ American Express;

Account Number: _____ Expiration date: _____

Signature: *X* _____

Name: _____ Institution: _____

Address: _____

        _____

City: _____ State: _____ Zip: _____

Phone Number: _____ FAX Number: _____

MAIL or *FAX* THIS ENTIRE ORDER FORM TO:

Haworth Document Delivery Service   **or FAX:** 1-800-895-0582
The Haworth Press, Inc.   **or CALL:** 1-800-429-6784
10 Alice Street   (9am-5pm EST)
Binghamton, NY 13904-1580

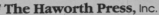

DATE DUE

JAN 2 6 2000	
MAY 1 0 2000	
JUN 0 8 2000	
AUG 1 9 2002	
AUG 1 8 2009	